ELECTRONICS BUYER'S GUIDE

3RD EDITION

BY EDWARD A. HALL

TABLE OF CONTENTS

FIRST EDITION

FOURTH PRINTING

Printed in the United States of America

Library of Congress Cataloging in Publication Data

Hall, Edward A.
Electronics buyers' guide.

1. Electronic apparatus and appliances—Directories.
2. Electronic apparatus and appliances—Catalogs.
I. Title.
TK7870.H225 1983 621.381'029'4 83-17972
ISBN 0-8306-0345-X (pbk.)

DIRECTIONS FOR USE

All of the components listed in this directory are available by mail order in small quantities from the firms listed. The following rules explain how information is listed in the directory:

1. Components are listed alphabetically by generic name as the first word of each line (e.g., CONNECTOR:, TRANSISTOR:, SOCKET:, etc.).
2. A specific part number is listed if the part is widely used or if it is a highly complex/special purpose part.
3. A general part number may be listed to refer to a broad class of many similar parts. An example of this is the SN7400 series of integrated circuits. Instead of individually listing each part in the series (such as the SN7432), they are lumped together into the "7400 series." They are then listed under "IC: 7400 SERIES, TTL." A series of parts may also be listed with "X" showing generality. For example, a 2N914 transistor would be included in "TRANSISTOR: 2NXXXX SERIES."
4. The number listed in the right-hand column of the directory gives the number of parts available in that series and is an indication of how complete the offering is from that supplier.
5. A general description is often given to further explain a part number or to distinguish the part if no part number is given.
6. The manufacturer of the part is often listed when they are well known for supplying that part.

When locating a component in the directory, start by finding the generic name listed alphabetically. Then look under both the part number (specific or general) and the description, because the component could be listed either way. Once you locate the component, the supplier can be evaluated by looking at the variety of components they offer, indicated by the number in the right-hand column. The address and phone number for the selected supplier can then be found in the Supplier Addresses section in the back of this guide.

The parts listed in the Directory are limited to those being sold by retail mail order. We are always looking for more retail mail-order suppliers, and this directory will become more complete as more suppliers are found. All suppliers who would like to be considered for inclusion in our directory should send us a listing or catalog of parts offered. Comments and suggestions about the directory are also welcome. Send them to Hallward Products, 39 Sunset Court, St. Louis, MO 63121.

NOTICE TO USERS

While every effort is extended to make the *Electronics Buyer's Guide* complete and accurate, the publisher makes no representation that it is absolutely accurate or complete. Errors and omissions, typographical, clerical, and otherwise sometimes do occur, and this possibility exists with respect to anything printed in this work. Hence, nothing within should be relied upon in any instance wherein there is a possibility of any loss or damage resulting from any statement, error, or omission in this volume.

TABLE OF ABBREVIATIONS

Abbreviation	Meaning
"	INCHES
'	FEET
>	GREATER THAN
A/D	ANALOG TO DIGITAL
AC	ALTERNATING CURRENT
AM	AMPLITUDE MODULATED
ANT	ANTENNA
APPL	APPLICATION
ASST	ASSORTMENT
ATV	AMATEUR TELEVISION
AUTO	AUTOMOBILE
AV	AVANTEK
BARR	BARRIER
BC	BROADCAST
BCD	BINARY CODED DECIMAL
BP	BAND PASS
CB	CITIZENS BAND
CKT	CIRCUIT
CLAD DS	COPPER COATED ON 2 SIDES
CLAD SS	COPPER COATED ON 1 SIDE
CLAD	COPPER COATED CIRCUIT
COMB HD	COMBINATION HEAD
COND	CONDUCTOR
CONT	CONTROLLED
CONTR	CONTROL
CRT	CATHODE RAY TUBE
CT	CENTER TAPPED
CW	CONTINUOUS WAVE
DBLE BAL	DOUBLE BALANCED
DBLE F	DOUBLE FEMALE
DC	DIRECT CURRENT
DEFL	DEFLECTION
DIP	DUAL IN-LINE PACKAGE
DS	SEE CLAD DS
DTMF	DUAL TONE MULTI-FREQUENCY (TELEPHONE STD)
EL	ELEMENT
ELECT	ELECTROLYTIC
EPROM	ERASABLE PROM
EXT	EXTERNAL
FET	FIELD EFFECT TRANSISTOR
FREQ	FREQUENCY
GEN R	GENERAL RADIO
GHZ	GIGAHERTZ
GLASS EP	GLASS EPOXY
Ga	GAUGE
HEW PA	HEWLETT PACKARD
HF	HIGH FREQUENCY (3 TO 40 MHZ)
HI-P	HIGH PASS
HI-REL	HIGH RELIABILITY
HP	HIGH POWER
HV	HIGH VOLTAGE (GREATER THAN 1000 VOLTS)
HY	HENRY
IC	INTEGRATED CIRCUIT
ID	INSIDE DIAMETER
IF	INTERMEDIATE FREQUENCY
INS	INSERT
INS	INSULATION
INT	INTERNAL
K	KHZ
kHz	KILOHERTZ
KV	KILOVOLT (1000 VOLTS)
L	LENGTH
LD	LEAD
LF	LOW FREQUENCY (30 TO 300 KHZ
LP	LOW POWER
LV	LOW VOLTAGE (LESS THAN 100 VOLTS)
LW-P	LOW PASS
M	MHZ
MACH	MACHINED
MDTL	MOTOROLA DIODE-TRANSISTOR LOGIC
MECL	MOTOROLA EMITTER COUPLED LOGIC
MF	MEDIUM FREQUENCY (30 KHZ TO 3 MHZ)
MHTL	MOTOROLA HIGH NOISE IMMUNITY LOGIC
MHZ	MEGAHERTZ
MM	MILLIMETER (1/1000 METER)
MOT	MOTOROLA
MPU	MICROPROCESSOR UNIT
MRTL	MOTOROLA RESISTOR-TRANSISTOR LOGIC
MT	MOUNT
MTTL	MOTOROLA TRANSISTOR-TRANSISTOR LOGIC
MULT	MULTIPLIER
MV	MEDIUM VOLTAGE

	(100 TO 1000 VOLTS)
MW	MILLIWATT (.001 WATT)
NP	NICKEL PLATED
OD	OUTER DIAMETER
OSC	OSCILLATOR
PB	PERFORATED BOARD
PC	PRINTED CIRCUIT
PCB	PRINTED CIRCUIT BOARD
PDR	POWDER
PIV	PEAK INVERSE VOLTAGE
PORC	PORCELAIN
POS	POSITION
PROM	PROGRAMMABLE READ ONLY MEMORY
PS	POWER SUPPLY
RAM	RANDOM ACCESS MEMORY
RCVR	RECEIVER
RECT	RECTIFIER
REG	REGULATOR
REPL	REPLACEMENT
RF	RADIO FREQUENCY
ROM	READ ONLY MEMORY
SCHOT	SCHOTTKY
SCR ADJ	SCREW DRIVER ADJUSTABLE
SENS	PHOTO SENSITIZED
SIP	SINGLE IN-LINE PACKAGE
SPRD	SPREADER
SS	SEE CLAD SS
SS	STAINLESS STEEL
STD	STANDARD
SUBMINI	SUBMINIATURE
SYNTH	SYNTHESIZER
T	THICK
TERM	TERMINATION
THRED	THREADED
THUMBWHL	THUMB WHEEL
TI	TEXAS INSTRUMENTS
TRK	TRACK
TTL	TRANSISTOR-TRANSISTOR LOGIC
TV	TELEVISION
TWT	TRAVELLING WAVE TUBE
U	USED
UHF	ULTRA HIGH FREQUENCY (300 TO 3000 MHZ)
UU	UN-USED
VAC	VOLTS ALTERNATING CURRENT
VAR	VARIABLE
VDC	VOLTS DIRECT CURRENT
VHF	VERY HIGH FREQUENCY (30 TO 300 MHZ)
VPS	VIBRATIONS PER SECOND
W	WATT
WB	WIDE BAND
WV	WORKING VOLTAGE
WW	WIRE WOUND
XFMR	TRANSFORMER
XMIT	TRANSMIT
XMTING	TRANSMITTING
XTAL	CRYSTAL

COMPONENT	COMPANY NAME	VARIETY STOCKED
ACTUATOR: LINEAR	Fair Radio Sales	2
ALARM: BEEPER	Semiconductors Surplus	2
ALARM: BELL	ETCO	8
ALARM: BELL, 24 V DC	Semiconductors Surplus	1
ALARM: BELL, 48 V DC	Digital Research	1
ALARM: BELL, 48 VDC	BCD Radio Parts	1
ALARM: BUZZER	ETCO	7
ALARM: BUZZER, 3-24 V DC, LARGE	Semiconductors Surplus	1
ALARM: BUZZER, 3-9 V DC, SMALL	Jameco Electronics	2
ALARM: BUZZER, 7 TO 17 VDC	Star-Tronics	1
ALARM: BUZZER, MINI, ELECTRONIC	Digi-Key	4
ALARM: DM03- BUZZER, SOLID STATE	Fuji-Svea	5
ALARM: HORN, 12 V	Surplus Electronics	1
ALARM: HORN, 18-24 V DC	Digital Research	1
ALARM: HORN, 6-12 V DC	Digital Research	1
ALARM: HORN, 6-24 VDC	BCD Radio Parts	2
ALARM: HORN, SIMPLEX, 6 V AC	Fair Radio Sales	1
ALARM: PIEZO, SOLID STATE	Mouser Electronics	10
ALARM: SC-12, SONALERT, MALLORY	Semiconductors Surplus	1
ALARM: SONALERT	Jameco Electronics	1
ALARM: SONALERT, MALLORY	Mouser Electronics	3
ALARM: TONE BUZZER, SOMA	Semiconductors Surplus	1
ALLIGATOR CLIP:	See Clip: Alligator	
AMPLIFIER: AK-1000M,500-1000 MHZ,AV	MHZ Electronics	1
AMPLIFIER: AL-45-0-1,450-800 MHZ,AV	MHZ Electronics	1
AMPLIFIER: AP-20-T, 200-400 MHZ, AV	MHZ Electronics	1
AMPLIFIER: AUDIO MODULE	ETCO	7
AMPLIFIER: AUDIO, 5.8 WATT	BCD Radio Parts	1
AMPLIFIER: AUDIO, 5.8 WATT	Digital Research	1
AMPLIFIER: AUDIO, 8 WATT, TO-220	BCD Radio Parts	1
AMPLIFIER: AUDIO, 8 WATT, TO-220	Digital Research	1
AMPLIFIER: CA602/CA, 15-270 MHZ	Semiconductors Surplus	1
AMPLIFIER: ESM- SERIES, AUDIO, PWR	Mouser Electronics	3
AMPLIFIER: HY- SERIES, AUDIO MODULE	Active Electronics	6
AMPLIFIER: KLYSTRON, 7.1-10 GHZ	Lectronic	1
AMPLIFIER: LOW NOISE,HF-UHF,AVANTEK	MHZ Electronics	4
AMPLIFIER: MC SERIES, RF, 5-900 MHZ	California Eastern Lab	10
AMPLIFIER: MHW 1172R, 40-300 MHZ	MHZ Electronics	1
AMPLIFIER: MHW 252, 25 W, 144 MHZ	Semiconductors Surplus	1
AMPLIFIER: MHW 401-2, 440-512 MHZ	Semiconductors Surplus	1
AMPLIFIER: MHW 591, 1-250 MHZ, MOT	Semiconductors Surplus	1
AMPLIFIER: MHW 592, 1-250 MHZ, MOT	Semiconductors Surplus	1
AMPLIFIER: MHW 612A, 20 W, 148 MHZ	Semiconductors Surplus	1
AMPLIFIER: MHW 612A, 20 W, 148 MHZ	MHZ Electronics	1
AMPLIFIER: MHW 613A, 30 W, 160 MHZ	MHZ Electronics	1
AMPLIFIER: MHW 710, 13W,400-512 MHZ	MHZ Electronics	1
AMPLIFIER: MHW 720, 20W,400-470 MHZ	MHZ Electronics	1
AMPLIFIER: MRF 450, 4 W IN,50 W OUT	Semiconductors Surplus	1
AMPLIFIER: PA5, 10 WATT, 435 MHz	P.C. Electronics	1
AMPLIFIER: RF BROADBAND, HF TO VHF	International Crystal	1
AMPLIFIER: RF POWER, TRANSISTOR	International Crystal	1
AMPLIFIER: RF, TRANSISTOR	International Crystal	
AMPLIFIER: T-44 MODULE, 432 MHZ	Amateur Radio Components Ser	1
AMPLIFIER: UTC2-102M,30-200 MHZ, AV	MHZ Electronics	1
AMPLIFIER: ZHL- , 1 W, .002-1 GHZ	Mini-Circuits Laboratory	3

COMPONENT	COMPANY NAME	VARIETY STOCKED
ANGLE STOCK: BRASS	Small Parts	5
ANGLE STOCK: STEEL	Small Parts	5
ANTENNA ROD:	See Inductor: Part Number	
ANTENNA: BOOM STOCK, ALUMINUM	Viking Instruments	2
ANTENNA: BOOM STOCK, FIBERGLASS	Viking Instruments	2
ANTENNA: CLAMP, BOOM TO EL, INSULTD	Viking Instruments	12
ANTENNA: CLAMP, BOOM TO ELEMENT	Viking Instruments	12
ANTENNA: CLAMP, TELESCOPING ELEMENT	Ham Hardware Headquarters	
ANTENNA: COIL, HF, SHORTENER	G & C Communications	2
ANTENNA: DIPOLE, MICROWAVE, ADJ	Lectronic	1
ANTENNA: GAMMA MATCH, COAX REACT	Viking Instruments	4
ANTENNA: INSULATOR	See Insulator:	
ANTENNA: MAST SECTIONS, TUBULAR	Fair Radio Sales	6
ANTENNA: MAST TO BOOM CENTER PLATE	Skylane Products	1
ANTENNA: MAST TO BOOM MOUNT	Viking Instruments	2
ANTENNA: QUAD ELEMENT DIA REDUCERS	Van Gorden Engineering	2
ANTENNA: QUAD RING TRANSFORMER	Skylane Products	1
ANTENNA: QUAD SPIDER	Van Gorden Engineering	1
ANTENNA: QUAD SPIDER HUB	Viking Instruments	5
ANTENNA: QUAD SPIDER HUB	Van Gorden Engineering	1
ANTENNA: QUAD SPIDER HUB	Skylane Products	1
ANTENNA: QUAD SPIDER STEM INSULATOR	Van Gorden Engineering	2
ANTENNA: QUAD SPIDER, BAMBOO	Skylane Products	2
ANTENNA: QUAD SPIDER, FIBERGLASS	Viking Instruments	6
ANTENNA: QUAD SPIDER, FIBERGLASS	Skylane Products	2
ANTENNA: QUAD SPIDER, FIBERGLASS	Alpha Electronic Labs	
ANTENNA: QUAD TERMINALS	Viking Instruments	1
ANTENNA: TELESCOPING, TV	Fuji-Svea	14
ANTENNA: TELESCOPING,PORTABLE RADIO	Surplus Electronics	1
ANTENNA: TELESCOPING,PORTABLE RADIO	Mouser Electronics	5
ANTENNA: TRAP, HF, WIRE	Barker & Williamson	2
ANTENNA: TRAP, HF, WIRE	Lacue Communications	2
ANTENNA: TRAP, HF, WIRE	Radiokit	2
ANTENNA: TRAP, HF, WIRE, W2VS	G & C Communications	1
ANTENNA: V-SUPPORT, DELTA LOOP	Viking Instruments	3
ANTENNA: WAVEGUIDE HORN, 10.525 GHZ	Alaska Microwave Lab	1
ANTENNA: WAVEGUIDE HORN, 6 GHZ	Star-Tronics	1
ANTENNA: WAVEGUIDE HORN,.75-170 GHZ	Lectronic	
ATTENUATOR: MICROWAVE, BOARD MOUNT	Elcom Systems	1
ATTENUATOR: MICROWAVE, COAX, 50 OHM	Elcom Systems	6
ATTENUATOR: MICROWAVE, COAX, 75 OHM	Elcom Systems	1
ATTENUATOR: MICROWAVE, COAX, 93 OHM	Elcom Systems	1
ATTENUATOR: MICROWAVE, TRIMMER	Elcom Systems	1
ATTENUATOR: RF, COAXIAL, 25-2000 W	Fuji-Svea	9
ATTENUATOR: RF, COAXIAL, FIXED	Lectronic	
ATTENUATOR: RF, COAXIAL, VARIABLE	Lectronic	
ATTENUATOR: RF, WAVEGUIDE, ADJ	Lectronic	
ATTENUATOR: RF, WAVEGUIDE, FIXED	Lectronic	
BALL JOINT: STEEL	Small Parts	10
BALLS: METAL, 1/16 TO 1" DIAMETER	Small Parts	67
BALLS: PLASTIC, 1/8 TO 1" DIAMETER	Small Parts	32
BAR STOCK: ALUMINUM	Small Parts	36
BAR STOCK: BRASS	Small Parts	48

COMPONENT	COMPANY NAME	VARIETY STOCKED
BAR STOCK: COPPER	Small Parts	13
BAR STOCK: PHOSPHOR BRONZE	Small Parts	9
BAR STOCK: STAINLESS STEEL	Small Parts	21
BAR STOCK: STEEL, COLD ROLLED	Small Parts	49
BATTERY CONNECTOR: 9 VOLT	Mouser Electronics	14
BATTERY CONNECTOR: 9 VOLT	Jameco Electronics	1
BATTERY CONNECTOR: 9 VOLT	ETCO	1
BATTERY CONNECTOR: 9 VOLT	Surplus Electronics	1
BATTERY CONNECTOR: 9 VOLT	Aldelco	1
BATTERY CONNECTOR: 9 VOLT	BCD Radio Parts	1
BATTERY CONNECTOR: FAHNESTOCK	Aldelco	1
BATTERY CONNECTOR: FAHNESTOCK	ETCO	1
BATTERY CONNECTOR: FAHNESTOCK	Mouser Electronics	2
BATTERY HOLDER: 9V, CLIP, STEEL	Mouser Electronics	2
BATTERY HOLDER: AA, C, D, ALUMINUM	Digi-Key	11
BATTERY HOLDER: AA,C,D, SNAP-IN	Aldelco	7
BATTERY HOLDER: AA,C,D,9V, SNAP-IN	Mouser Electronics	72
BATTERY HOLDER: AAA, A, D CELL	ETCO	9
BATTERY HOLDER: C CELL	Jameco Electronics	2
BATTERY HOLDER: C CELL	Star-Tronics	2
BATTERY HOLDER: CLIP, NO CONTACTS	ETCO	16
BATTERY: 9 VOLT	Mouser Electronics	1
BATTERY: ALKALINE, 1.5 & 9 VOLT	Mouser Electronics	5
BATTERY: ALKALINE, AA, C, D	Digi-Key	3
BATTERY: CARBON ZINC, AA, C, D, 9V	Digi-Key	4
BATTERY: GEL-TYPE, RECHARGABLE	Active Electronics	9
BATTERY: GEL-TYPE, RECHARGABLE	Mouser Electronics	8
BATTERY: GEL-TYPE, RECHARGABLE	Surplus Electronics	2
BATTERY: GELLED ELECTO LEAD ACID	Digi-Key	5
BATTERY: LEAD CELL, DRY CHARGE	Fair Radio Sales	2
BATTERY: LEAD-ACID, 6 & 12 V	Surplus Electronics	5
BATTERY: LEAD-ACID, RECHARGABLE,12V	Fair Radio Sales	1
BATTERY: LITHIUM, CALCULATOR, ECT.	Digi-Key	10
BATTERY: MERCURY, 1.25 VOLT	Surplus Electronics	1
BATTERY: NICAD	Star-Tronics	3
BATTERY: NICAD	Digi-Key	18
BATTERY: NICAD	Active Electronics	13
BATTERY: NICAD	Amateur Electronic Supply	6
BATTERY: NICAD	Semiconductors Surplus	8
BATTERY: NICAD	ETCO	6
BATTERY: NICAD, 1.2 & 6V, RECONDND	Edmund Scientific	3
BATTERY: NICAD, 12V DC, 1.2 AH	Fair Radio Sales	1
BATTERY: NICAD, AA & 9 VOLT	Surplus Electronics	2
BATTERY: NICAD, AA, C, D	G & C Communications	3
BATTERY: NICAD, AA, C, D	Mouser Electronics	3
BATTERY: NICAD, AA, GENERAL ELECT	MHZ Electronics	1
BATTERY: NICAD, AAA MINI PENLIGHT	ETCO	1
BATTERY: NICAD, LARGE	Fair Radio Sales	2
BATTERY: WATER ACTIVATED, 7.5&114 V	Fair Radio Sales	2
BATTERY: ZINC CARBON, 1.5 & 9 VOLT	Mouser Electronics	5
BEARING: FLANGE, SINTERED BRONZE	Small Parts	16
BEARING: NEEDLE ROLLER	Small Parts	12
BEARING: PRESS, ALIGNABLE	Small Parts	6
ARING: SLEEVE, SINTERED BRONZE	Small Parts	16
ARING: THRUST, SINTERED BRONZE	Small Parts	8

COMPONENT	COMPANY NAME	VARIETY STOCKED
BELT: "O" RING, 2.75-3.875" ID	ETCO	4
BELT: FLAT, 6" CIRCUMFERENCE,3/16"W	ETCO	1
BELT: FLAT, 6-16" CIRCUMFERENCE	Fuji-Svea	20
BELT: FLAT, 9" CIRCUMFERENCE, 1/4"W	ETCO	1
BELT: ROUND, 6-18.5" CIRCUMFERENCE	Fuji-Svea	19
BELT: ROUND, ELASTIC, ENDLESS	Small Parts	3
BELT: SQUARE, 2.5-10" CIRCUMFERENCE	Fuji-Svea	11
BELT: TOOTHED, TIMING	Small Parts	21
BEZEL FILTER: RED, 2.25 X .5"	Digital Research	1
BEZEL FILTER: RED, 2.25 X .5"	BCD Radio Parts	1
BEZEL: LED DISPLAY	Mouser Electronics	15
BEZEL: LED DISPLAY	Quest Electronics	6
BOARD: COPPER CLAD & PERFERATED	See Circuit Board:	
BOLOMETER: MICROWAVE	Lectronic	
BULB:	See Lamp:	
BUS STRIP: CIRCUIT BOARD	Quest Electronics	1
BUSS BAR:	See Also Bar Stock	
BUSS BAR: 0.665 X 0.05", COPPER	Electric Motion Company	1
CABINET BAIL: TILT STAND	Daytapro Electronics	3
CABINET:	See Also Case:	
CABINET:	See Also Chassis:	
CABINET: ALUMINUM	Radiokit	4
CABINET: ALUMINUM	Daytapro Electronics	53
CABINET: ALUMINUM, BUILT IN CHASSIS	Mouser Electronics	3
CABINET: ALUMINUM, BUILT IN CHASSIS	Circuit Specialists	6
CABINET: ALUMINUM, CHASSIS/COVER	Digi-Key	9
CABINET: ALUMINUM, COLOR STYLED,LMB	Mouser Electronics	4
CABINET: ALUMINUM, CROWN ROYAL	Circuit Specialists	25
CABINET: ALUMINUM, LMB	Circuit Specialists	63
CABINET: DESK CONSOLE, ALUMINUM	Mouser Electronics	10
CABINET: DESK CONSOLE, ALUMINUM	Daytapro Electronics	12
CABINET: DESK CONSOLE, ALUMINUM	Circuit Specialists	7
CABINET: DESK CONSOLE, ALUMINUM	BCD Radio Parts	2
CABINET: DESK CONSOLE, ALUMINUM	Digi-Key	15
CABINET: DESK CONSOLE, ALUMINUM,LMB	Mouser Electronics	6
CABINET: DESK CONSOLE, EQUIPMENT	Sintec	3
CABINET: DESK CONSOLE, PLASTIC	Mouser Electronics	6
CABINET: DESK CONSOLE, PLASTIC	Active Electronics	6
CABINET: DESK CONSOLE, PLASTIC	Daytapro Electronics	5
CABINET: EQUIPMENT	Ten-Tec	32
CABINET: EQUIPMENT, PLASTIC	Active Electronics	33
CABINET: KEYBOARD	Daytapro Electronics	1
CABINET: KEYBOARD, PLASTIC	Quest Electronics	1
CABINET: LIFT TOP, STEEL, LARGE	Fair Radio Sales	1
CABINET: LOW SILHOUETTE	Daytapro Electronics	8
CABINET: LOW SILHOUETTE, TEN-TEC	Circuit Specialists	12
CABINET: METAL CHASSIS, TEN-TEC	BCD Radio Parts	1
CABINET: MFJ- SERIES, EQUIPMENT	MJF Enterprises	6
CABINET: PERFORATED, ALUMINUM, LMB	Mouser Electronics	3

COMPONENT	COMPANY NAME	VARIETY STOCKED
CABINET: RACK PANEL, STANDARD 19"	Radiokit	7
CABINET: RACK, STANDARD 19", STEEL	Fair Radio Sales	2
CABINET: RACK, STANDARD 19", STEEL	Atlantic Surplus Sales	1
CABINET: RACK-PANEL WIDTH, PORTABLE	Ten-Tec	13
CABINET: SLOPING FRONT	Ten-Tec	12
CABINET: SLOPING FRONT, MOD-U-BOX	Circuit Specialists	5
CABINET: SLOPING FRONT, MODULAR	Radiokit	10
CABLE CLAMP: ADHESIVE MOUNTING	Van Gorden Engineering	1
CABLE CLAMP: ADHESIVE MT, KWIK-KLIP	Small Parts	10
CABLE CLAMP: NAIL MOUNT	Aldelco	1
CABLE CLAMP: SCREW MOUNT	Surplus Electronics	2
CABLE CLAMP: SCREW MOUNT	Mouser Electronics	11
CABLE CLAMP: SCREW MOUNT, NYLON	Small Parts	9
CABLE CLAMP: STEEL	Mouser Electronics	4
CABLE CLAMP: STEEL AND RUBBER	Star-Tronics	1
CABLE TIE MOUNT: ADHESIVE	Small Parts	2
CABLE TIE MOUNT: ADHESIVE	Mouser Electronics	1
CABLE TIE:	Daytapro Electronics	4
CABLE TIE:	Digital Research	2
CABLE TIE:	BCD Radio Parts	1
CABLE TIE:	Small Parts	3
CABLE TIE:	Semiconductors Surplus	2
CABLE TIE: BEADED	Surplus Electronics	1
CABLE TIE: NYLON	Mouser Electronics	9
CABLE TIE: NYLON	Fuji-Svea	5
CABLE TIE: PANDUIT, RED, 15" L	Star-Tronics	1
CABLE TIE: RELEASABLE, NYLON	Small Parts	2
CABLE TIE: TWIST LOCK	Mouser Electronics	1
CABLE TIE: TWIST LOCK, NYLON	Small Parts	4
CABLE TIE: WITH IDENTIFICATION PLT	Small Parts	1
CABLE: AC CORD, 2-CONDUCTOR	Mouser Electronics	8
CABLE: AC CORD, 2-CONDUCTOR	Fuji-Svea	1
CABLE: AC CORD, 2-CONDUCTOR	Surplus Electronics	1
CABLE: AC CORD, 2-CONDUCTOR	Aldelco	1
CABLE: AC CORD, 2-CONDUCTOR	Semiconductors Surplus	1
CABLE: AC CORD, 2-CONDUCTOR	ETCO	2
CABLE: AC CORD, 2-CONDUCTOR	Jameco Electronics	5
CABLE: AC CORD, 3-COND, NO PLUG	Surplus Electronics	1
CABLE: AC CORD, 3-COND, NO PLUG	Mouser Electronics	1
CABLE: AC CORD, 3-CONDUCTOR	BCD Radio Parts	1
CABLE: AC CORD, 3-CONDUCTOR	Surplus Electronics	1
CABLE: AC CORD, 3-CONDUCTOR	Mouser Electronics	19
CABLE: AC CORD, 3-CONDUCTOR	Fair Radio Sales	2
CABLE: AC CORD, 3-CONDUCTOR	Jameco Electronics	4
CABLE: AC CORD, 3-CONDUCTOR, 14 Ga	Amp Supply	1
CABLE: AC CORD, BUSINESS MACH PLUG	Fair Radio Sales	1
CABLE: AC CORD, DETACHABLE, 3-COND	Mouser Electronics	6
CABLE: AC CORD, EUROPEAN, 3-COND	Mouser Electronics	1
CABLE: AC CORD, MOLDED RECPT, HEW PA	Surplus Electronics	1
CABLE: AC CORD, TV CHEATER	Aldelco	2
CABLE: ANTENNA ROTOR	Amateur Electronic Supply	4
CABLE: ANTENNA ROTOR	Lacue Communications	2
CABLE: ANTENNA ROTOR	Certified International	
CABLE: ANTENNA ROTOR	Nemal Electronics	1
CABLE: ANTENNA ROTOR	Semiconductors Surplus	1

COMPONENT	COMPANY NAME	VARIETY STOCKED
CABLE: ANTENNA ROTOR, 8 CONDUCTOR	Alpha Electronic Labs	1
CABLE: AUDIO COAX	Nemal Electronics	1
CABLE: AUDIO PATCH	Surplus Electronics	3
CABLE: AUTO CIGAR LIGHTER PLUG/CORD	ETCO	1
CABLE: BRAID, COPPER, FLEXIBLE	Electric Motion Company	1
CABLE: COAX	See Cable: RG-#	
CABLE: COAX, PREPARED	Amateur Electronic Supply	23
CABLE: COMPUTER CONTROL, 15 COND	Fair Radio Sales	1
CABLE: CONNECTOR, WITH PLUG	Electronic Marketplace	9
CABLE: DIP JUMPER	See Cable: Jumper, Dip	
CABLE: JUMPER, CARD EDGE CONNECTOR	Quest Electronics	15
CABLE: JUMPER, CARD EDGE CONNECTOR	Digi-Key	56
CABLE: JUMPER, DIP SOCKET	Mouser Electronics	8
CABLE: JUMPER, DIP SOCKET	Circuit Specialists	49
CABLE: JUMPER, DIP SOCKET	Sintec	138
CABLE: JUMPER, DIP SOCKET	Quest Electronics	20
CABLE: JUMPER, DIP SOCKET	Digi-Key	107
CABLE: JUMPER, DIP SOCKET	Fuji-Svea	21
CABLE: JUMPER, DIP SOCKET	Fair Radio Sales	10
CABLE: JUMPER, DIP SOCKET	Jameco Electronics	24
CABLE: JUMPER, DIP SOCKET	Active Electronics	105
CABLE: JUMPER, PCB AND DP25 ENDS	Surplus Electronics	1
CABLE: JUMPER, PCB CONNECTOR	Digi-Key	55
CABLE: JUMPER, PCB CONNECTOR	Quest Electronics	10
CABLE: JUMPER, PT-TO-PT, STRIPPED	Sintec	5
CABLE: JUMPER, SINGLE ROW PLUG	Sintec	48
CABLE: JUMPER, SOCKET CONNECTOR	Digi-Key	55
CABLE: JUMPER, SOCKET CONNECTOR	Quest Electronics	25
CABLE: JUMPER, SOCKET CONNECTOR	Jameco Electronics	8
CABLE: LADDER LINE, 450 OHM	Certified International	
CABLE: LADDER LINE, 450 OHM	Lacue Communications	
CABLE: MICROPHONE	Certified International	
CABLE: MICROPHONE, COILED	Mouser Electronics	3
CABLE: MICROPHONE, COILED	Fair Radio Sales	4
CABLE: MICROPHONE, COILED	Surplus Electronics	4
CABLE: MICROPHONE, COILED	Amateur Electronic Supply	1
CABLE: MICROPHONE, DOUBLE SHEILD	Star-Tronics	1
CABLE: MICROPHONE,TAPE RECORDER,STD	ETCO	2
CABLE: MODEM ASSEMBLY	Surplus Electronics	2
CABLE: MULTICONDUCTOR	Digi-Key	22
CABLE: MULTICONDUCTOR, FOIL SHEILD	Digi-Key	10
CABLE: MULTICONDUCTOR, TWISTED PAIR	Digi-Key	14
CABLE: OPEN LINE, 450 OHM, POLY INS	Lacue Communications	
CABLE: PHONO, 1/4" PLUG, 21" CORD	BCD Radio Parts	1
CABLE: PHONO, ADAPTER	BCD Radio Parts	4
CABLE: POWER, AUTO STEREO,UNIVERSAL	ETCO	2
CABLE: POWER, LV DC, POLARIZED PLUG	ETCO	1
CABLE: POWER, POLARIZED CONNECTORS	Aldelco	1
CABLE: POWER, POLARIZED CONNECTORS	Fuji-Svea	2
CABLE: POWER, TAPE PLAYER,UNIVERSAL	Fuji-Svea	2
CABLE: RG-11 72 OHM SOLID, FOAM	Amateur Electronic Supply	
CABLE: RG-11 72 OHM STRANDED, COAX	Amateur Electronic Supply	
CABLE: RG-115U TEFLON, SILVER	Nemal Electronics	
CABLE: RG-11AU MIL SPEC, NON-CONTAM	Nemal Electronics	
CABLE: RG-11U 72 OHM FOAM COAX	Fair Radio Sales	
CABLE: RG-11U 72 OHM FOAM COAX	Semiconductors Surplus	
CABLE: RG-11U 72 OHM FOAM COAX	Lacue Communications	
CABLE: RG-11U 75 OHM, 80% SHEILD	Nemal Electronics	
CABLE: RG-11U 75 OHM, MIL SPEC	Nemal Electronics	

COMPONENT	COMPANY NAME	VARIETY STOCKED
CABLE: RG-141U TEFLON INSUL, COAX	Radiokit	
CABLE: RG-142BU COAX	Radiokit	
CABLE: RG-142U DBL SILVER, TEFLON	Nemal Electronics	
CABLE: RG-174	Alpha Electronic Labs	
CABLE: RG-174 STRANDED, POLY	Certified International	
CABLE: RG-174 STRANDED, POLY	Semiconductors Surplus	
CABLE: RG-174U 50 OHM, MIL SPEC	Nemal Electronics	
CABLE: RG-174U COAX	Fox Tango Corporation	
CABLE: RG-174U COAX	Radiokit	
CABLE: RG-213 MIL SPEC	Lacue Communications	
CABLE: RG-213 NON-CONTAMINATING	Certified International	
CABLE: RG-213 NON-CONTAMINATING	Amateur Electronic Supply	
CABLE: RG-213U NON-CONT, MIL SPEC	Nemal Electronics	
CABLE: RG-214 DOUBLE SHEILD RG-8	Alpha Electronic Labs	
CABLE: RG-214U DOUBLE SILVER SHEILD	Nemal Electronics	
CABLE: RG-217U 50 OHM, DBL SHEILD	Nemal Electronics	
CABLE: RG-223 DOUBLE SHEILD	Alpha Electronic Labs	
CABLE: RG-223 DOUBLE SHIELD, POLY	Certified International	
CABLE: RG-223U 50 OHM DOUBLE SHEILD	Nemal Electronics	
CABLE: RG-393U TEFLON, SILVER, 12"L	Nemal Electronics	
CABLE: RG-55BU DOUBLE SHEILD 50 OHM	Nemal Electronics	
CABLE: RG-58 SOLID COAX	Amateur Electronic Supply	
CABLE: RG-58 STRANDED, POLY COAX	Certified International	
CABLE: RG-58A STRANDED, FOAM COAX	Amateur Electronic Supply	
CABLE: RG-58A/U FOAM, COAX	G & C Communications	
CABLE: RG-58A/U STRANDED, MIL SPEC	Nemal Electronics	
CABLE: RG-58C NON-CONTAMINATING	Amateur Electronic Supply	
CABLE: RG-58C/U NON-CONT, MIL SPEC	Nemal Electronics	
CABLE: RG-58U	Fuji-Svea	
CABLE: RG-58U	Semiconductors Surplus	
CABLE: RG-58U 53.5 OHM	Fair Radio Sales	
CABLE: RG-58U 80% SHEILD	Nemal Electronics	
CABLE: RG-58U 96% SHEILD, MIL SPEC	Nemal Electronics	
CABLE: RG-58U MIL SPEC	Lacue Communications	
CABLE: RG-59 72 OHM SOLID COAX	Amateur Electronic Supply	
CABLE: RG-59 72 OHM STRANDED, POLY	Certified International	
CABLE: RG-59U	Fuji-Svea	
CABLE: RG-59U	ETCO	
CABLE: RG-59U 72 OHM STRANDED, COAX	Fair Radio Sales	
CABLE: RG-59U 72 OHM STRANDED, COAX	Semiconductors Surplus	
CABLE: RG-59U 72 OHM STRANDED, COAX	Lacue Communications	
CABLE: RG-59U 72 OHM, 100% FOIL	Nemal Electronics	
CABLE: RG-59U 96% SHEILD, MIL SPEC	Nemal Electronics	
CABLE: RG-62 STRANDED, 93 OHM COAX	Certified International	
CABLE: RG-62 STRANDED, POLY, COAX	Certified International	
CABLE: RG-62A/U 93 OHM, MIL SPEC	Nemal Electronics	
CABLE: RG-6A/U 75 OHM, DBL SHEILD	Nemal Electronics	
CABLE: RG-6U 75 OHM, 100% SHEILD	Nemal Electronics	
CABLE: RG-8 COAX	Barker & Williamson	1
CABLE: RG-8 DOUBLE SHEILD, COAX	Certified International	
CABLE: RG-8 SOLID DIELECTRIC, COAX	G & C Communications	
CABLE: RG-8 STRANDED, COAX	Amateur Electronic Supply	
CABLE: RG-8 STRANDED, FOAM COAX	Certified International	
CABLE: RG-8 STRANDED, FOAM COAX	Amateur Electronic Supply	
CABLE: RG-8 STRANDED, VINYL COAX	Certified International	
CABLE: RG-8A NON-CONTAMINATING	Amateur Electronic Supply	
CABLE: RG-8U	Alpha Electronic Labs	
CABLE: RG-8U 80% SHEILD	Nemal Electronics	
CABLE: RG-8U 95% SHEILD	Nemal Electronics	

COMPONENT	COMPANY NAME	VARIETY STOCKED
CABLE: RG-8U 96% SHEILD, MIL SPEC	Nemal Electronics	
CABLE: RG-8U 97% SHEILD, 11 Ga	Nemal Electronics	
CABLE: RG-8U FOAM, COAX	G & C Communications	
CABLE: RG-8U STRANDED, FOAM COAX	Lacue Communications	
CABLE: RG-8U STRANDED, FOAM COAX	Fair Radio Sales	
CABLE: RG-8X	Alpha Electronic Labs	
CABLE: RG-8X MINI EIGHT, 95% SHEILD	Nemal Electronics	
CABLE: RG-8X MINI EIGHT, COAX	G & C Communications	
CABLE: RG-8X STRANDED, FOAM COAX	Amateur Electronic Supply	
CABLE: RG-8X STRANDED, FOAM COAX	Lacue Communications	
CABLE: RG-8X STRANDED, FOAM COAX	Certified International	
CABLE: RIBBON	Jameco Electronics	5
CABLE: RIBBON	Semiconductors Surplus	6
CABLE: RIBBON	Quest Electronics	2
CABLE: RIBBON, COLOR CODED	Mouser Electronics	6
CABLE: RIBBON, COLOR CODED	Digi-Key	11
CABLE: RIBBON, COLOR CODED	Active Electronics	13
CABLE: RIBBON, COLOR CODED	Electronic Marketplace	1
CABLE: RIBBON, GREY	Active Electronics	13
CABLE: RIBBON, WITH GROUND PLANE	Star-Tronics	1
CABLE: RIGID COAX, 50 OHM, 1/4x28"	Star-Tronics	1
CABLE: ROTOR, ANTENNA	See Cable: Antenna Rotor	
CABLE: RS-232 WITH CONNECTORS	Digi-Key	88
CABLE: RS-232 WITH CONNECTORS	Quest Electronics	15
CABLE: SEMI-RIGID, .141" DIAMETER	Alaska Microwave Lab	
CABLE: SEMI-RIGID, .141" DIAMETER	RIW Products	
CABLE: STAINLESS STEEL, .024-1/8"	Small Parts	5
CABLE: TELEPHONE, 3-CONDUCTOR	ETCO	1
CABLE: TWIN LEAD 300 OHM, 1KW	Amateur Electronic Supply	
CABLE: TWIN LEAD 300 OHM, FOAM	ETCO	
CABLE: TWIN LEAD 72 OHM, 1 KW	Amateur Electronic Supply	
CABLE: TWIN LEAD 72 OHM, 100 WATT	Amateur Electronic Supply	
CABLE: TWISTED, 3 CONDUCTOR, TEFLON	Star-Tronics	1
CALORIMETER: MICROWAVE	Lectronic	
CAM: LINEAR RISE, FAST FALL, 1/4"MT	Small Parts	6
CAPACITOR CLAMP: 1 3/8 TO 1 7/16"	ETCO	1
CAPACITOR: 15,000 WV, .05uF, MYLAR	Star-Tronics	1
CAPACITOR: 20-150,000 uF, PS	Fair Radio Sales	40
CAPACITOR: 22-110,000 uF, MV	Semiconductors Surplus	12
CAPACITOR: 400 TO 7500 WV	Fair Radio Sales	24
CAPACITOR: BATHTUB CONDENSER TYPE	Fair Radio Sales	10
CAPACITOR: CERAMIC	BCD Radio Parts	9
CAPACITOR: CERAMIC, 1.6-10 KV	Semiconductors Surplus	8
CAPACITOR: CERAMIC, 5 KV	Amp Supply	4
CAPACITOR: CERAMIC, 5.1-50 pF	Semiconductors Surplus	
CAPACITOR: CERAMIC, LEADLESS	Alaska Microwave Lab	1
CAPACITOR: CERAMIC, MILITARY CK-	Active Electronics	38
CAPACITOR: CERAMIC, MONOLYTHIC, MV	Babylon Electronics	3
CAPACITOR: CERAMIC, MONOLYTHIC	Proverty Electronics	1
CAPACITOR: CERAMIC, MONOLYTHIC, LV	Active Electronics	18
CAPACITOR: CERAMIC, MONOLYTHIC, MV	Active Electronics	28
CAPACITOR: CERAMIC, TUBULAR, 500 WV	Digital Research	3
CAPACITOR: CHIP, MICROWAVE	MHZ Electronics	60
CAPACITOR: CHIP, MICROWAVE	Alaska Microwave Lab	28
CAPACITOR: CHIP, MICROWAVE	Semiconductors Surplus	48

COMPONENT	COMPANY NAME	VARIETY STOCKED
CAPACITOR: COMPUTER GRADE	Quest Electronics	22
CAPACITOR: COMPUTER GRADE	Surplus Electronics	25
CAPACITOR: DISC, 1000 WV	Fair Radio Sales	3
CAPACITOR: DISC, 3000 WV	Fair Radio Sales	2
CAPACITOR: DISC, CERAMIC	Digital Research	1
CAPACITOR: DISC, CERAMIC	Daytapro Electronics	34
CAPACITOR: DISC, CERAMIC	Alpha Electronic Labs	
CAPACITOR: DISC, CERAMIC	Radiokit	
CAPACITOR: DISC, CERAMIC, 12-500 WV	Mouser Electronics	179
CAPACITOR: DISC, CERAMIC, 6000 WV	Star-Tronics	7
CAPACITOR: DISC, CERAMIC, HV	ETCO	4
CAPACITOR: DISC, CERAMIC, HV	Key Electronics	1
CAPACITOR: DISC, CERAMIC, HV	Active Electronics	2
CAPACITOR: DISC, CERAMIC, LV	ETCO	2
CAPACITOR: DISC, CERAMIC, LV	Key Electronics	2
CAPACITOR: DISC, CERAMIC, LV	Active Electronics	7
CAPACITOR: DISC, CERAMIC, LV	Proverty Electronics	5
CAPACITOR: DISC, CERAMIC, LV	Star-Tronics	4
CAPACITOR: DISC, CERAMIC, LV	Circuit Specialists	19
CAPACITOR: DISC, CERAMIC, LV	Jameco Electronics	17
CAPACITOR: DISC, CERAMIC, LV	Quest Electronics	62
CAPACITOR: DISC, CERAMIC, MV	Babylon Electronics	16
CAPACITOR: DISC, CERAMIC, MV	Key Electronics	6
CAPACITOR: DISC, CERAMIC, MV	Active Electronics	25
CAPACITOR: DISC, CERAMIC, MV	Star-Tronics	20
CAPACITOR: DISC, CERAMIC, MV	ETCO	4
CAPACITOR: DISC, NPO TYPE	See Capacitor: NPO	
CAPACITOR: DOOR KNOB, 5 KV	Radiokit	3
CAPACITOR: DOOR KNOB, 5-15 KV	Semiconductors Surplus	5
CAPACITOR: ELECTROLYTIC	See Also Capacitor: Multi-Se	
CAPACITOR: ELECTROLYTIC, ALUMINUM	Babylon Electronics	15
CAPACITOR: ELECTROLYTIC, ALUMINUM	Fuji-Svea	156
CAPACITOR: ELECTROLYTIC, ALUMINUM	Daytapro Electronics	16
CAPACITOR: ELECTROLYTIC, ALUMINUM	Alpha Electronic Labs	
CAPACITOR: ELECTROLYTIC, CAN, MV	Fuji-Svea	17
CAPACITOR: ELECTROLYTIC, CAN-TYPE	Mouser Electronics	2
CAPACITOR: ELECTROLYTIC, HIGH FREQ	Digi-Key	21
CAPACITOR: ELECTROLYTIC, HV	Surplus Electronics	2
CAPACITOR: ELECTROLYTIC, LV	Digi-Key	349
CAPACITOR: ELECTROLYTIC, LV	Surplus Electronics	57
CAPACITOR: ELECTROLYTIC, LV	Star-Tronics	24
CAPACITOR: ELECTROLYTIC, LV	ETCO	100
CAPACITOR: ELECTROLYTIC, LV	Active Electronics	23
CAPACITOR: ELECTROLYTIC, LV	Digital Research	6
CAPACITOR: ELECTROLYTIC, LV	Radiokit	20
CAPACITOR: ELECTROLYTIC, LV	Proverty Electronics	8
CAPACITOR: ELECTROLYTIC, LV	BCD Radio Parts	27
CAPACITOR: ELECTROLYTIC, MINI, LV	Key Electronics	6
CAPACITOR: ELECTROLYTIC, MINI, LV	Mouser Electronics	336
CAPACITOR: ELECTROLYTIC, MINI, LV	Jameco Electronics	47
CAPACITOR: ELECTROLYTIC, MINI, LV	Sintec	43
CAPACITOR: ELECTROLYTIC, MINI, LV	Circuit Specialists	47
CAPACITOR: ELECTROLYTIC, MINI, LV	Quest Electronics	41
CAPACITOR: ELECTROLYTIC, MV	BCD Radio Parts	6
CAPACITOR: ELECTROLYTIC, MV	ETCO	18
CAPACITOR: ELECTROLYTIC, MV	Mouser Electronics	66
CAPACITOR: ELECTROLYTIC, MV	Fuji-Svea	5
CAPACITOR: ELECTROLYTIC, MV	Amp Supply	1
CAPACITOR: ELECTROLYTIC, MV	Circuit Specialists	54

COMPONENT	COMPANY NAME	VARIETY STOCKED
CAPACITOR: ELECTROLYTIC, MV	Active Electronics	5
CAPACITOR: ELECTROLYTIC, MV	Digital Electronics	1
CAPACITOR: ELECTROLYTIC, RADIAL LD	Circuit Specialists	36
CAPACITOR: ELECTROLYTIC,.01uF,7500V	Fair Radio Sales	1
CAPACITOR: ERIE 1201-785, .1-10 GHZ	Semiconductors Surplus	1
CAPACITOR: ERIE 1270-016, .2-10 GHZ	Semiconductors Surplus	1
CAPACITOR: FEED THRU	ETCO	1
CAPACITOR: FEED THRU	Alaska Microwave Lab	3
CAPACITOR: FEED THRU, CERAMIC	Radiokit	1
CAPACITOR: FEED THRU, SOLDER TYPE	Semiconductors Surplus	8
CAPACITOR: FEED THRU, STUD MOUNT	Semiconductors Surplus	3
CAPACITOR: GLASS PISTON	Fair Radio Sales	3
CAPACITOR: MICA	Daytapro Electronics	12
CAPACITOR: MICA, ARCO ELEMENCO	Circuit Specialists	57
CAPACITOR: MICA, HV	Amp Supply	12
CAPACITOR: MICA, JAN CM- SIZE, HV	Fair Radio Sales	8
CAPACITOR: MICA, MV	Star-Tronics	2
CAPACITOR: MICA, RF, F2L SIZE	Fair Radio Sales	
CAPACITOR: MICA, SILVER DIPPED	ETCO	27
CAPACITOR: MICA, SILVER DIPPED	Jameco Electronics	18
CAPACITOR: MICA, SILVER DIPPED	Mouser Electronics	33
CAPACITOR: MICA, SILVER DIPPED	Radiokit	FULL
CAPACITOR: MICA, SILVER DIPPED	Alpha Electronic Labs	
CAPACITOR: MICA, XMTING, 1-35 KV	Fair Radio Sales	34
CAPACITOR: MULTI-SECTION, ELECT, MV	ETCO	3
CAPACITOR: MULTI-SECTION, ELECT, MV	Surplus Electronics	1
CAPACITOR: MULTI-SECTION, ELECT, MV	Fuji-Svea	55
CAPACITOR: MULTI-SECTION, ELECT, MV	Fair Radio Sales	11
CAPACITOR: MULTI-SECTION, ELECT, MV	Star-Tronics	2
CAPACITOR: MYLAR	Fair Radio Sales	2
CAPACITOR: MYLAR	Daytapro Electronics	29
CAPACITOR: MYLAR	Quest Electronics	26
CAPACITOR: MYLAR	Alpha Electronic Labs	
CAPACITOR: MYLAR, LV	ETCO	8
CAPACITOR: MYLAR, LV	Surplus Electronics	14
CAPACITOR: MYLAR, MV	Digital Research	3
CAPACITOR: MYLAR, MV	ETCO	21
CAPACITOR: MYLAR, MV	Jameco Electronics	13
CAPACITOR: MYLAR, MV	Babylon Electronics	6
CAPACITOR: NON-POLARIZED, 1-50 uF	Mouser Electronics	23
CAPACITOR: NON-POLARIZED, 2-2500 uF	ETCO	8
CAPACITOR: NON-POLARIZED,.47-470 uF	Fuji-Svea	30
CAPACITOR: NPO, TEMPERATURE CONTR	Circuit Specialists	77
CAPACITOR: NPO, TEMPERATURE CONTR	Digi-Key	48
CAPACITOR: OIL FILLED,53.3mF,3.5 KV	Semiconductors Surplus	1
CAPACITOR: ORANGE DROP, TUBULAR, HV	Fuji-Svea	25
CAPACITOR: ORANGE DROP, TUBULAR, MV	Fuji-Svea	46
CAPACITOR: ORANGE DROP, TUBULAR, MV	Circuit Specialists	130
CAPACITOR: PAPER, .1uf, 2 KV, ELPAC	Semiconductors Surplus	1
CAPACITOR: PAPER, 6 IN DECADE SET	Star-Tronics	1
CAPACITOR: PAPER, HV	Star-Tronics	2
CAPACITOR: PAPER, MV	Star-Tronics	18
CAPACITOR: PHOTO-FLASH	Fair Radio Sales	1
CAPACITOR: PHOTO-FLASH	Surplus Electronics	3
CAPACITOR: PHOTO-FLASH, RUBYCON	Mouser Electronics	2
CAPACITOR: POLYCARBONATE, MV	ETCO	1
CAPACITOR: POLYCON POLYESTER, MV	Proverty Electronics	1
CAPACITOR: POLYESTER FILM MYLAR	Circuit Specialists	11
CAPACITOR: POLYESTER FILM MYLAR, MV	Key Electronics	8

COMPONENT	COMPANY NAME	VARIETY STOCKED
CAPACITOR: POLYESTER FILM, METFILM	Fuji-Svea	9
CAPACITOR: POLYESTER FILM, MV	Mouser Electronics	147
CAPACITOR: POLYESTER FILM, MV	ETCO	13
CAPACITOR: POLYESTER FILM, MV	Star-Tronics	15
CAPACITOR: POLYESTER FILM, MV	Fuji-Svea	34
CAPACITOR: POLYESTER FILM, MV	ETCO	6
CAPACITOR: POLYESTER FILM, MV	Digi-Key	153
CAPACITOR: POLYESTER FILM, PACER	Fuji-Svea	9
CAPACITOR: POLYESTER FILM, VERTI-	Fuji-Svea	11
CAPACITOR: POLYESTER, MV	Star-Tronics	6
CAPACITOR: POLYPROPYLENE FILM	ETCO	65
CAPACITOR: POLYPROPYLENE FILM	Mouser Electronics	17
CAPACITOR: POLYPROPYLENE FILM	Circuit Specialists	19
CAPACITOR: POLYSTRENE	Fair Radio Sales	2
CAPACITOR: POLYSTYRENE	Alpha Electronic Labs	
CAPACITOR: POLYSTYRENE	Daytapro Electronics	20
CAPACITOR: POLYSTYRENE FILM, MV	Mouser Electronics	35
CAPACITOR: POLYSTYRENE, MV	ETCO	1
CAPACITOR: PRECISION, .0511 uF	Star-Tronics	1
CAPACITOR: SANGHMD TYPE 2A	Semiconductors Surplus	1
CAPACITOR: SILVER DIPPED MICA	See Capacitor: Mica	
CAPACITOR: TANTALUM	Daytapro Electronics	17
CAPACITOR: TANTALUM	Quest Electronics	26
CAPACITOR: TANTALUM	ETCO	17
CAPACITOR: TANTALUM	Radiokit	8
CAPACITOR: TANTALUM	Alpha Electronic Labs	
CAPACITOR: TANTALUM REPLACEMENT	Digi-Key	28
CAPACITOR: TANTALUM, DIPPED	Semiconductors Surplus	1
CAPACITOR: TANTALUM, DIPPED	Jameco Electronics	21
CAPACITOR: TANTALUM, DIPPED	Active Electronics	25
CAPACITOR: TANTALUM, DIPPED	Mouser Electronics	105
CAPACITOR: TANTALUM, DIPPED	Proverty Electronics	1
CAPACITOR: TANTALUM, DIPPED	Key Electronics	4
CAPACITOR: TANTALUM, DIPPED	Circuit Specialists	11
CAPACITOR: TANTALUM, DIPPED	Babylon Electronics	12
CAPACITOR: TANTALUM, DIPPED	Sintec	23
CAPACITOR: TANTALUM, DIPPED, RESIN	Fuji-Svea	59
CAPACITOR: TANTALUM, DIPPED, RESIN	Digi-Key	78
CAPACITOR: TANTALUM, DIPPED, SOLID	Aldelco	15
CAPACITOR: TANTALUM, LV	Star-Tronics	17
CAPACITOR: TANTALUM, MOLDED	Surplus Electronics	7
CAPACITOR: TANTALUM, MV, 150 WV	Star-Tronics	1
CAPACITOR: TANTALUM, PCB-TYPE	Fair Radio Sales	6
CAPACITOR: TANTALUM, SINTERED	Mouser Electronics	18
CAPACITOR: TUBULAR, MV	ETCO	14
CAPACITOR: UNELCO, 6.8-1000pF	Semiconductors Surplus	16
CAPACITOR: VACUUM, 3 pF, 15 KV	Fair Radio Sales	3
CAPACITOR: VARIABLE, 2 SECTION	ETCO	3
CAPACITOR: VARIABLE, 2 SECTION	Caywood Electronics	17
CAPACITOR: VARIABLE, 2 SECTION	BCD Radio Parts	1
CAPACITOR: VARIABLE, 2 SECTION	Radiokit	29
CAPACITOR: VARIABLE, 2 SECTION	Digital Research	1
CAPACITOR: VARIABLE, 2 SECTION	D & V Radio	7
CAPACITOR: VARIABLE, 2 SECTION, BAL	Caywood Electronics	36
CAPACITOR: VARIABLE, 3 SECTION	Radiokit	2
CAPACITOR: VARIABLE, AIR	Alpha Electronic Labs	
CAPACITOR: VARIABLE, AIR	Radiokit	8
CAPACITOR: VARIABLE, AIR	Star-Tronics	1
CAPACITOR: VARIABLE, AIR	ETCO	20

COMPONENT	COMPANY NAME	VARIETY STOCKED
CAPACITOR: VARIABLE, AIR	Semiconductors Surplus	19
CAPACITOR: VARIABLE, AIR TRIMMER	Fair Radio Sales	2
CAPACITOR: VARIABLE, AIR, ALL STAR	Surplus Electronics	1
CAPACITOR: VARIABLE, AIR, CARDWELL	D & V Radio	10
CAPACITOR: VARIABLE, AIR, CARDWELL	Radiokit	18
CAPACITOR: VARIABLE, AIR, HAMMARLND	D & V Radio	12
CAPACITOR: VARIABLE, AIR, JOHNSON	Mouser Electronics	15
CAPACITOR: VARIABLE, AIR, JOHNSON	D & V Radio	5
CAPACITOR: VARIABLE, AIR, JOHNSON	Circuit Specialists	20
CAPACITOR: VARIABLE, AIR, JOHNSON	Radiokit	17
CAPACITOR: VARIABLE, AIR, MILLEN	Radiokit	90
CAPACITOR: VARIABLE, AIR, MILLEN	Caywood Electronics	58
CAPACITOR: VARIABLE, AIR, MILLEN	D & V Radio	11
CAPACITOR: VARIABLE, AIR, PADDER	Radiokit	4
CAPACITOR: VARIABLE, AIR, SCR ADJ	D & V Radio	11
CAPACITOR: VARIABLE, ARCO	Semiconductors Surplus	15
CAPACITOR: VARIABLE, ARCO	Alpha Electronic Labs	
CAPACITOR: VARIABLE, BROADCAST RCVR	Circuit Specialists	1
CAPACITOR: VARIABLE, BROADCAST RCVR	ETCO	9
CAPACITOR: VARIABLE, BROADCAST RCVR	Star-Tronics	1
CAPACITOR: VARIABLE, BROADCAST RCVR	Mouser Electronics	7
CAPACITOR: VARIABLE, BROADCAST RCVR	Radiokit	1
CAPACITOR: VARIABLE, BUTTERFLY, AIR	Radiokit	6
CAPACITOR: VARIABLE, BUTTERFLY, AIR	D & V Radio	5
CAPACITOR: VARIABLE, BUTTERFLY, HP	Barker & Williamson	2
CAPACITOR: VARIABLE, CERAMIC	Alpha Electronic Labs	
CAPACITOR: VARIABLE, CERAMIC, MINI	Semiconductors Surplus	8
CAPACITOR: VARIABLE, CERAMIC, TRIM	Mouser Electronics	33
CAPACITOR: VARIABLE, CERAMIC, TRIM	Circuit Specialists	5
CAPACITOR: VARIABLE, CERAMIC, TRIM	Key Electronics	3
CAPACITOR: VARIABLE, CERAMIC, TRIM	Jameco Electronics	5
CAPACITOR: VARIABLE, CERAMIC, TRIM	Quest Electronics	22
CAPACITOR: VARIABLE, CERAMIC, TRIM	ETCO	5
CAPACITOR: VARIABLE, COMMAND RCVR	Fair Radio Sales	1
CAPACITOR: VARIABLE, DIFFERENTIAL	D & V Radio	6
CAPACITOR: VARIABLE, DIFFERENTIAL	Caywood Electronics	2
CAPACITOR: VARIABLE, DIFFERENTIAL	Radiokit	4
CAPACITOR: VARIABLE, DOUBLE END	Caywood Electronics	44
CAPACITOR: VARIABLE, DOUBLE END	D & V Radio	2
CAPACITOR: VARIABLE, DOUBLE END	Radiokit	3
CAPACITOR: VARIABLE, FM TUNING	Radiokit	2
CAPACITOR: VARIABLE, HAMMARLUND	Radiokit	31
CAPACITOR: VARIABLE, HIGH VOLTAGE	See Capacitor: Variable, Tra	
CAPACITOR: VARIABLE, MICA, COMPRESS	Circuit Specialists	38
CAPACITOR: VARIABLE, MICA, COMPRESS	Radiokit	26
CAPACITOR: VARIABLE, MICA, COMPRESS	Star-Tronics	1
CAPACITOR: VARIABLE, MICA, COMPRESS	ETCO	13
CAPACITOR: VARIABLE, MICA, SCR ADJ	Fuji-Svea	9
CAPACITOR: VARIABLE, MINI, SCR ADJ	Circuit Specialists	2
CAPACITOR: VARIABLE, OFFSET PLATES	D & V Radio	3
CAPACITOR: VARIABLE, PC, SCR ADJ	Surplus Electronics	2
CAPACITOR: VARIABLE, PC, SCR ADJ	Mouser Electronics	9
CAPACITOR: VARIABLE, PISTON TRIMMER	Alaska Microwave Lab	3
CAPACITOR: VARIABLE, PISTON TRIMMER	Key Electronics	1
CAPACITOR: VARIABLE, PISTON TRIMMER	Alpha Electronic Labs	
CAPACITOR: VARIABLE, PISTON TRIMMER	Semiconductors Surplus	20
CAPACITOR: VARIABLE, POLYPROPYLENE	Semiconductors Surplus	5
CAPACITOR: VARIABLE, RECEIVER	Fair Radio Sales	22
CAPACITOR: VARIABLE, RECEIVER	Star-Tronics	1

COMPONENT	COMPANY NAME	VARIETY STOCKED
CAPACITOR: VARIABLE, SOLID DIELECT	Circuit Specialists	3
CAPACITOR: VARIABLE, STEATITE	Fuji-Svea	5
CAPACITOR: VARIABLE, STEATITE, TRIM	Fuji-Svea	2
CAPACITOR: VARIABLE, TEFLON TRIMMER	Mouser Electronics	2
CAPACITOR: VARIABLE, TRANSMITTING	Fair Radio Sales	21
CAPACITOR: VARIABLE, TRANSMITTING	Caywood Electronics	13
CAPACITOR: VARIABLE, TRANSMITTING	ETCO	6
CAPACITOR: VARIABLE, TRANSMITTING	Amp Supply	6
CAPACITOR: VARIABLE, TRANSMITTING	Radiokit	16
CAPACITOR: VARIABLE, TRIMMER	BCD Radio Parts	3
CARTRIDGE NEEDLE: RECORD PLAYER	ETCO	6
CARTRIDGE: RECORD PLAYER	ETCO	5
CASE:	See Also Cabinet:	
CASE: ADJUSTA- BOX, TILT BRACKET	Mouser Electronics	2
CASE: ALUMINUM	Semiconductors Surplus	6
CASE: ALUMINUM	Radiokit	10
CASE: ALUMINUM, INTERLOCKING, LMB	Radiokit	25
CASE: BAKELITE	Semiconductors Surplus	2
CASE: CLOCK, FRONT BEZEL	BCD Radio Parts	1
CASE: CLOCK, FRONT BEZEL	Digital Research	1
CASE: CLOCK, FRONT BEZEL	Digi-Key	1
CASE: CLOCK, FRONT BEZEL, BLACK	Semiconductors Surplus	1
CASE: CLOCK/INSTRUMENT, FRONT BEZEL	Jameco Electronics	1
CASE: DESK TOP, SLOPING FRONT	Jameco Electronics	3
CASE: DESK TOP, VECTOR	Mouser Electronics	8
CASE: DIECAST ALUMINUM BOX	Radiokit	11
CASE: DIECAST ALUMINUM BOX, HAMMOND	Mouser Electronics	5
CASE: EQUIPMENT, CARRYING HANDLE	Active Electronics	10
CASE: EQUIPMENT, CARRYING, PLASTIC	Active Electronics	5
CASE: HAND-HELD SHAPE, INSTRUMENT	Sintec	1
CASE: HAND-HELD SHAPE, INSTRUMENT	Jameco Electronics	1
CASE: INSTRUMENT	Digital Research	3
CASE: INSTRUMENT, ALUMINUM PANEL	Active Electronics	2
CASE: INSTRUMENT, PLASTIC	Active Electronics	2
CASE: INSTRUMENT, PLASTIC, AL PANEL	BCD Radio Parts	2
CASE: INSTRUMENT, SLOPING, PLASTIC	Sintec	2
CASE: INSTRUMENT, SLOPING, PLASTIC	Mouser Electronics	2
CASE: INSTRUMENT, SLOPING, PLASTIC	Jameco Electronics	2
CASE: MOLDED PLASTIC, OPEN FACE	BCD Radio Parts	1
CASE: PHENOLIC, INTEGRAL FEET	BCD Radio Parts	1
CASE: PHENOLIC, WITH PANEL	Digi-Key	6
CASE: PHENOLIC, WITH PANEL, BLACK	Small Parts	8
CASE: PHENOLIC, WITH PANEL, BLACK	ETCO	5
CASE: PLASTIC, ALUMINUM PANEL	Mouser Electronics	3
CASE: PLASTIC, BENCHTOP, METAL ENDS	Jameco Electronics	2
CASE: PLASTIC, RED CLEAR FRONT	Jameco Electronics	1
CASE: PLASTIC, WITH PANEL	Daytapro Electronics	8
CASE: PLASTIC, WITH PANEL, HAMMOND	Mouser Electronics	25
CASE: PLASTIC, WITH PANEL, HAMMOND	Circuit Specialists	5
CASE: PLUG-IN, COMPONENT, OCTAL PLG	ETCO	1
CASE: PROBE SHAPE, THREADED TIP	Jameco Electronics	1
CASE: STEEL BOX, MINIATURE, LMB	Mouser Electronics	7
CASE: TONE PAD HOUSING, 12 HOLE	Aldelco	1
CASE: UTILITY BOX	ETCO	2
CASE: UTILITY BOX, SLOPING, METAL	Mouser Electronics	1
CASE: UTILITY BOX, STEEL	Mouser Electronics	4

COMPONENT	COMPANY NAME	VARIETY STOCKED
CASE: UTILITY, SLOPING FRONT	Daytapro Electronics	11
CASE: VIDEO MONITOR, METAL, CAST	Semiconductors Surplus	1
CASSETTE HEAD:	See Head:	
CAVITY: RESONANT	See Resonator: Cavity	
CHAIN: BEAD, STAINLESS STEEL	Small Parts	1
CHAIN: ROLLER LINK, 1/4", PLASTIC	Small Parts	1
CHAIN: ROLLER LINK, 1/4", SS	Small Parts	5
CHANNEL STOCK: BRASS	Small Parts	5
CHASSIS: ALUMINUM	Daytapro Electronics	18
CHASSIS: ALUMINUM BOX	Digi-Key	22
CHASSIS: ALUMINUM, HAMMOND	Radiokit	18
CHASSIS: JIFFY BOX, ALUMINUM, LMB	Mouser Electronics	13
CHASSIS: TITE-FIT BOX	Daytapro Electronics	21
CHASSIS: TITE-FIT BOX, LMB	Radiokit	19
CHASSIS: TITE-FIT BOX, LMB	Mouser Electronics	32
CHEMICAL: ANTI-STATIC SPRAY	Digi-Key	1
CHEMICAL: CIRCUIT BOARD CLEANER	Kepro Circuit Systems	2
CHEMICAL: CIRCUIT BOARD ETCH	ETCO	2
CHEMICAL: CIRCUIT BOARD ETCH	Kepro Circuit Systems	2
CHEMICAL: CIRCUIT BOARD ETCH	Active Electronics	1
CHEMICAL: CIRCUIT COOLER	Mouser Electronics	1
CHEMICAL: CIRCUIT COOLER	Fuji-Svea	1
CHEMICAL: CIRCUIT COOLER, SPRAY	Active Electronics	1
CHEMICAL: CLEANER, SPRAY-KLEEN	Active Electronics	1
CHEMICAL: CONTACT CLEANER	Mouser Electronics	3
CHEMICAL: CONTACT CLEANER	Digi-Key	2
CHEMICAL: COOLER SPRAY, DRI-FREEZE	Digi-Key	1
CHEMICAL: COPPER ELECTROLITE	Kepro Circuit Systems	2
CHEMICAL: CUTTING FLUID	Small Parts	2
CHEMICAL: DEGREASER, DRI-KLEENER	Digi-Key	1
CHEMICAL: DIAL CHORD "NO SLIP"	ETCO	1
CHEMICAL: ETCH DEVELOPING SOLUTION	Active Electronics	1
CHEMICAL: ETCH DEVELOPING SOLUTION	Mouser Electronics	2
CHEMICAL: ETCH RESIST SENSITIZER	Mouser Electronics	3
CHEMICAL: ETCH RESIST, PEN	Active Electronics	2
CHEMICAL: ETCH REVERSING DEVELOPER	ETCO	1
CHEMICAL: FLUX REMOVER	Digi-Key	1
CHEMICAL: GLUE, INSTANT STICK	Fuji-Svea	1
CHEMICAL: GOLD PLATING SOLUTION	Kepro Circuit Systems	1
CHEMICAL: GREASE, FLEXIBLE SHAFT	Small Parts	1
CHEMICAL: HEAT SINK COMPOUND	See Chemical: Thermal Comp	
CHEMICAL: HELIUM GAS	Edmund Scientific	1
CHEMICAL: INSULATING VARNISH	Mouser Electronics	2
CHEMICAL: INSULATING VARNISH, GLPT	Active Electronics	1
CHEMICAL: LUBRICANT, PENETRATING	Digi-Key	1
CHEMICAL: LUMINOUS PAINT	Edmund Scientific	1
CHEMICAL: MAGNET CEMENT	Mouser Electronics	1
CHEMICAL: MOLDING COMPOUND, LATEX	Edmund Scientific	2
CHEMICAL: NICKEL ELECTROLITE	Kepro Circuit Systems	1
CHEMICAL: OIL, MOTOR	Small Parts	1
CHEMICAL: PHOTO RESIST DEVELOPER	Kepro Circuit Systems	6
CHEMICAL: PHOTO RESIST DEVELOPER	ETCO	2
CHEMICAL: PHOTO RESIST MARKING PEN	Kepro Circuit Systems	1

COMPONENT	COMPANY NAME	VARIETY STOCKED
CHEMICAL: PHOTO RESIST STRIPPER	Kepro Circuit Systems	4
CHEMICAL: PHOTO RESIST, CIRCUIT BRD	ETCO	2
CHEMICAL: PHOTO RESIST, CIRCUIT BRD	Kepro Circuit Systems	2
CHEMICAL: PLASTIC PATCH, FILLER	Fuji-Svea	1
CHEMICAL: PLASTIC/GLASS CLEANER	Digi-Key	1
CHEMICAL: POLISH, METAL & PLASTIC	Fuji-Svea	1
CHEMICAL: POLYURETHANE SPRAY	ETCO	1
CHEMICAL: PUTTY, HIGH VOLTAGE	Fuji-Svea	1
CHEMICAL: RESIN FLUX REMOVER	Mouser Electronics	1
CHEMICAL: RUST INHIBITOR	Digi-Key	1
CHEMICAL: SOLDER ELECTROLITE	Kepro Circuit Systems	1
CHEMICAL: SOLDER STRIPPER	Kepro Circuit Systems	1
CHEMICAL: SOLVENT, LACQUER	ETCO	1
CHEMICAL: TAPE HEAD CLEANER	Mouser Electronics	1
CHEMICAL: TAPE HEAD CLEANER, SPRAY	ETCO	1
CHEMICAL: TELESCOPING ANTENNA LUBE	ETCO	1
CHEMICAL: THERMAL COMPOUND	Semiconductors Surplus	1
CHEMICAL: THERMAL COMPOUND	Fuji-Svea	1
CHEMICAL: THERMAL COMPOUND	Active Electronics	3
CHEMICAL: THERMAL COMPOUND	Mouser Electronics	1
CHEMICAL: THREADING ADHESIVE	Small Parts	6
CHEMICAL: TIN PLATING, IMMERSION	Kepro Circuit Systems	1
CHEMICAL: TINNING SOLUTION, CKT BRD	Active Electronics	1
CHEMICAL: TUNER CLEANER	Fuji-Svea	3
CHEMICAL: TUNERLUB	Mouser Electronics	1
CIRCUIT BOARD EJECTOR:	See Hardware: Ejector	
CIRCUIT BOARD GUIDE:	See Hardware: Guide	
CIRCUIT BOARD HANDLE: PULLER	Daytapro Electronics	1
CIRCUIT BOARD HOLDER: 22.5x15x8"	Atlantic Surplus Sales	1
CIRCUIT BOARD: CHEMICALS	See Chemical:	
CIRCUIT BOARD: CLAD DS	Surplus Electronics	1
CIRCUIT BOARD: CLAD, DS	Active Electronics	3
CIRCUIT BOARD: CLAD, ETCH SENS, DS	Active Electronics	1
CIRCUIT BOARD: CLAD, ETCH SENS, SS	Active Electronics	1
CIRCUIT BOARD: CLAD, SS	Active Electronics	3
CIRCUIT BOARD: CLAD,ETCH SENSITIZED	Active Electronics	1
CIRCUIT BOARD: CLAD,ETCH SENSITIZED	Mouser Electronics	3
CIRCUIT BOARD: DIP PLUG, ETCHED	Active Electronics	5
CIRCUIT BOARD: DRILLED, BARE	Circuit Specialists	5
CIRCUIT BOARD: DRILLED, CLAD	Circuit Specialists	2
CIRCUIT BOARD: DRILLED, CLAD,VECTOR	Active Electronics	8
CIRCUIT BOARD: EPOXY, CLAD SS	Kepro Circuit Systems	8
CIRCUIT BOARD: EPOXY, CLAD, SENS	Kepro Circuit Systems	10
CIRCUIT BOARD: ETCHED, DIP PLUG	Circuit Specialists	2
CIRCUIT BOARD: ETCHED, DRILLED	Circuit Specialists	1
CIRCUIT BOARD: ETCHED, DRILLED	Jameco Electronics	7
CIRCUIT BOARD: ETCHED, EDGE CONNECT	Sintec	3
CIRCUIT BOARD: ETCHED, LED DISPLAY	BCD Radio Parts	1
CIRCUIT BOARD: ETCHED, LED DISPLAY	Digital Research	1.
CIRCUIT BOARD: ETCHED, UNIVERSAL	Active Electronics	5
CIRCUIT BOARD: ETCHED, UNIVERSAL	Daytapro Electronics	3
CIRCUIT BOARD: ETCHED, UNIVERSAL	Sintec	4
CIRCUIT BOARD: ETCHED, UNIVERSAL	Digi-Key	8
CIRCUIT BOARD: ETCHED, UNIVERSAL	ETCO	1

COMPONENT	COMPANY NAME	VARIETY STOCKED
CIRCUIT BOARD: ETCHED, UNIVERSAL	Star-Tronics	1
CIRCUIT BOARD: ETCHED, WIRE WRAP	Quest Electronics	7
CIRCUIT BOARD: ETCHED, WIRE WRAP	BCD Radio Parts	1
CIRCUIT BOARD: ETCHED,PROJECT BOARD	Dynaclad Industries	600
CIRCUIT BOARD: ETCHING, CUSTOM	Dynaclad Industries	
CIRCUIT BOARD: FIBERGLASS, CLAD DS	Star-Tronics	1
CIRCUIT BOARD: FIBERGLASS, CLAD DS	Fair Radio Sales	2
CIRCUIT BOARD: FIBERGLASS, CLAD SS	Fair Radio Sales	4
CIRCUIT BOARD: GLASS EP, CLAD, SENS	Kepro Circuit Systems	34
CIRCUIT BOARD: GLASS EPOXY, CLAD DS	Kepro Circuit Systems	3
CIRCUIT BOARD: GLASS EPOXY, CLAD DS	ETCO	2
CIRCUIT BOARD: GLASS EPOXY, CLAD SS	Kepro Circuit Systems	31
CIRCUIT BOARD: GLASS EPOXY, CLAD SS	ETCO	1
CIRCUIT BOARD: MICROWAVE	See Circuit Board: Teflon	
CIRCUIT BOARD: PAPER EPOXY, CLAD	Mouser Electronics	3
CIRCUIT BOARD: PERFORATED, BARE	Circuit Specialists	3
CIRCUIT BOARD: PERFORATED, BARE	Jameco Electronics	10
CIRCUIT BOARD: PERFORATED, BARE	Sintec	1
CIRCUIT BOARD: PERFORATED, BARE	ETCO	3
CIRCUIT BOARD: PERFORATED, BARE	Digi-Key	9
CIRCUIT BOARD: PERFORATED, BARE	Mouser Electronics	20
CIRCUIT BOARD: PERFORATED, CLAD	Mouser Electronics	4
CIRCUIT BOARD: PERFORATED, CLAD SS	Jameco Electronics	1
CIRCUIT BOARD: PHENOLIC, CLAD SS	Fair Radio Sales	3
CIRCUIT BOARD: PHENOLIC, CLAD SS	ETCO	3
CIRCUIT BOARD: PHENOLIC, CLAD SS	Kepro Circuit Systems	5
CIRCUIT BOARD: PHENOLIC, CLAD, SENS	Kepro Circuit Systems	6
CIRCUIT BOARD: S100, WIRE WRAP,CLAD	Quest Electronics	6
CIRCUIT BOARD: TEFLON, CLAD DS	Alaska Microwave Lab	3
CIRCUIT BOARD: WIRE WRAP, 44 PIN	Quest Electronics	7
CIRCUIT BOARD: WIRE WRAP, 72 PIN	Quest Electronics	3
CIRCUIT BREAKER:	Fair Radio Sales	18
CIRCUIT BREAKER:	Surplus Electronics	7
CIRCUIT BREAKER:	Fuji-Svea	6
CIRCULATOR: COAXIAL, .1-8 GHZ	Lectronic	
CIRCULATOR: WAVEGUIDE, 2.6-75 GHZ	Lectronic	
CLIP: ALLIGATOR	Mouser Electronics	42
CLIP: ALLIGATOR	ETCO	1
CLIP: ALLIGATOR, INSULATED	ETCO	7
CLIP: ALLIGATOR, INSULATED, HI-REL	Mouser Electronics	1
CLIP: DOLPHIN, MINIATURE,SMOOTH JAW	ETCO	1
CLIP: FAHNESTOCK	See: Battery Connector	
CLIP: INDUCTOR TAP	Radiokit	1
CLIP: INDUCTOR TAP	Barker & Williamson	1
CLIP: METAL, SPRING, SCREW MOUNT	Small Parts	4
CLIP: MICROPHONE HOLDER, MAGNETIC	ETCO	1
CLIP: MICROPHONE HOLDER, MAGNETIC	Fuji-Svea	1
CLOCK MODULE:	Digi-Key	6
CLOCK MODULE:	ETCO	2
CLOCK MODULE:	BCD Radio Parts	1
CLOCK MODULE:	Jameco Electronics	8
CLOCK MOVEMENT: BATTERY, QUARTZ	ETCO	4
CLUTCH: ROLLER, DIRECTIONAL	Small Parts	4

COMPONENT	COMPANY NAME	VARIETY STOCKED
CLUTCH: SLIP, MINIATURE, ADJUSTABLE	Small Parts	2
COAX CABLE:	See Cable:	
COAX SEAL:	Lacue Communications	1
COAX SEAL:	BCD Radio Parts	1
COAX SEAL:	Radiokit	1
COAX SWITCH:	See Relay: Coax	
COIL FORM:	See Inductor Form:	
CONNECTOR: 735201/FX3850NF,TAPERLOC	Semiconductors Surplus	1
CONNECTOR: 78-880-1, RF, 1,PRODELIN	Semiconductors Surplus	1
CONNECTOR: 83- SERIES, COAX	G & C Communications	3
CONNECTOR: AC ELECTRICAL PLUG	Fuji-Svea	1
CONNECTOR: AC ELECTRICAL PLUG	Mouser Electronics	7
CONNECTOR: AC ELECTRICAL, PANEL MT	Daytapro Electronics	1
CONNECTOR: AC ELECTRICAL, PANEL MT	Mouser Electronics	15
CONNECTOR: ADAPTER, MOTOROLA TO UHF	Radiokit	1
CONNECTOR: ADAPTER, PHONE & PHONO	Mouser Electronics	14
CONNECTOR: ADAPTER, RF, COAXIAL	Lectronic	
CONNECTOR: ALLIGATOR CLIP	See Clip: Alligator	
CONNECTOR: AMPHENOL FULL LINE	Certified International	
CONNECTOR: AUDIO ADAPTER	Aldelco	10
CONNECTOR: AUTO ANTENNA JACK	Surplus Electronics	1
CONNECTOR: AUTO ANTENNA JACK, PLUG	ETCO	2
CONNECTOR: AUTO ANTENNA JACK, PLUG	Mouser Electronics	2
CONNECTOR: AUTO ANTENNA PLUG	Fuji-Svea	1
CONNECTOR: AUTO CIGAR LIGHTER	Mouser Electronics	2
CONNECTOR: AUTO CIGAR LIGHTER JACK	BCD Radio Parts	1
CONNECTOR: AUTO CIGAR LIGHTER JACK	Digital Research	1
CONNECTOR: AUTO CIGAR LIGHTER PLUG	Nemal Electronics	1
CONNECTOR: AUTO CIGAR LIGHTER PLUG	ETCO	1
CONNECTOR: AUTO CIGAR LIGHTER PLUG	Digital Research	1
CONNECTOR: AUTO CIGAR LIGHTER PLUG	Fuji-Svea	1
CONNECTOR: AUTO POWER DC JACK	Mouser Electronics	6
CONNECTOR: AUTO STEREO PLUG/JACK	ETCO	2
CONNECTOR: BANANA, HI-REL	Mouser Electronics	4
CONNECTOR: BANANA, INLINE SOCKET	ETCO	1
CONNECTOR: BANANA, JACK	ETCO	3
CONNECTOR: BANANA, JACK	Digi-Key	7
CONNECTOR: BANANA, JACK	Mouser Electronics	8
CONNECTOR: BANANA, JACK, 4 MM	ETCO	2
CONNECTOR: BANANA, PLUG	Mouser Electronics	6
CONNECTOR: BANANA, PLUG, DUAL,GEN R	Star-Tronics	1
CONNECTOR: BANANA, PLUG, SOLDER	Digi-Key	6
CONNECTOR: BANANA, PLUG, SOLDERLESS	Digi-Key	7
CONNECTOR: BINDING POST, 5 WAY	ETCO	8
CONNECTOR: BINDING POST, 5 WAY	Star-Tronics	1
CONNECTOR: BINDING POST, BANANA	Jameco Electronics	ASST
CONNECTOR: BINDING POST, BANANA	Mouser Electronics	12
CONNECTOR: BINDING POST, BANANA	Digi-Key	2
CONNECTOR: BINDING POST, INS MT	Sintec	1
CONNECTOR: BNC SERIES ADAPTER, COAX	Nemal Electronics	5
CONNECTOR: BNC SERIES, COAX	BCD Radio Parts	1
CONNECTOR: BNC SERIES, COAX	Surplus Electronics	11
CONNECTOR: BNC SERIES, COAX	Radiokit	2
CONNECTOR: BNC SERIES, COAX	Alpha Electronic Labs	1

COMPONENT	COMPANY NAME	VARIETY STOCKED
CONNECTOR: BNC SERIES, COAX	Daytapro Electronics	1
CONNECTOR: BNC SERIES, COAX	Jameco Electronics	5
CONNECTOR: BNC SERIES, COAX	Nemal Electronics	5
CONNECTOR: BNC SERIES, COAX	Alaska Microwave Lab	2
CONNECTOR: BNC SERIES, COAX	ETCO	9
CONNECTOR: BNC SERIES, COAX	Fair Radio Sales	14
CONNECTOR: BNC SERIES, COAX	Semiconductors Surplus	11
CONNECTOR: BNC SERIES, COAX	Aldelco	12
CONNECTOR: BNC TO SO-239 ADAPTER	G & C Communications	1
CONNECTOR: C- SERIES, COAX	Fair Radio Sales	4
CONNECTOR: CATV SERIES	Mouser Electronics	4
CONNECTOR: CENTRONICS, MALE	Semiconductors Surplus	1
CONNECTOR: CH- SERIES, COAX	Fair Radio Sales	1
CONNECTOR: CIRCUIT BOARD EDGE	Surplus Electronics	7
CONNECTOR: CIRCUIT BOARD EDGE	Alpha Electronic Labs	1
CONNECTOR: CIRCUIT BOARD EDGE	BCD Radio Parts	5
CONNECTOR: CIRCUIT BOARD EDGE	Mouser Electronics	12
CONNECTOR: CIRCUIT BOARD EDGE	Digi-Key	76
CONNECTOR: CIRCUIT BOARD EDGE	Quest Electronics	5
CONNECTOR: CIRCUIT BOARD EDGE	Jameco Electronics	1
CONNECTOR: CIRCUIT BOARD EDGE	Active Electronics	72
CONNECTOR: CIRCUIT BOARD EDGE	Mouser Electronics	6
CONNECTOR: CIRCUIT BOARD EDGE	ETCO	3
CONNECTOR: CIRCUIT BOARD EDGE	Jameco Electronics	6
CONNECTOR: CIRCUIT BOARD EDGE	Daytapro Electronics	5
CONNECTOR: CIRCUIT BOARD EDGE, TI	Sintec	34
CONNECTOR: CKT BOARD EDGE, 86 PIN	Semiconductors Surplus	1
CONNECTOR: D-TYPE	See Also Connector: RS-232	
CONNECTOR: D-TYPE, ACCESSORY PIN	Active Electronics	2
CONNECTOR: D-TYPE, ACCESSORY SOCKET	Active Electronics	2
CONNECTOR: D-TYPE, FEMALE SCREWLOCK	Active Electronics	1
CONNECTOR: D-TYPE, HOOD	Digi-Key	5
CONNECTOR: D-TYPE, SCREW RETAINER	Active Electronics	1
CONNECTOR: D-TYPE, SUBMINI, FEMALE	Active Electronics	4
CONNECTOR: D-TYPE, SUBMINI, FEMALE	Sintec	7
CONNECTOR: D-TYPE, SUBMINI, FEMALE	Digi-Key	5
CONNECTOR: D-TYPE, SUBMINI, FEMALE	Alpha Electronic Labs	4
CONNECTOR: D-TYPE, SUBMINI, FEMALE	Mouser Electronics	6
CONNECTOR: D-TYPE, SUBMINI, MALE	Mouser Electronics	6
CONNECTOR: D-TYPE, SUBMINI, MALE	Digi-Key	5
CONNECTOR: D-TYPE, SUBMINI, MALE	Active Electronics	4
CONNECTOR: D-TYPE, SUBMINI, MALE	Sintec	7
CONNECTOR: D-TYPE, SUBMINI, MALE	Alpha Electronic Labs	4
CONNECTOR: DA-113 DUMMY	Fair Radio Sales	1
CONNECTOR: DB25, SOCKET, PLUG	Surplus Electronics	2
CONNECTOR: DB25, SOCKET, PLUG	Electronic Marketplace	2
CONNECTOR: DB25P, SUBMINIATURE	Semiconductors Surplus	1
CONNECTOR: DC POWER JACK	Mouser Electronics	9
CONNECTOR: DC POWER JACK, RIGHT ANG	Digital Research	1
CONNECTOR: DC POWER PLUG	Mouser Electronics	4
CONNECTOR: DIN JACK	Mouser Electronics	6
CONNECTOR: DIN JACK, 5 PIN	Semiconductors Surplus	1
CONNECTOR: DIN PLUG	Mouser Electronics	4
CONNECTOR: DIN PLUG, 5 PIN	Semiconductors Surplus	1
CONNECTOR: DIN TYPE, HI-REL	Mouser Electronics	12
CONNECTOR: DIN TYPE, JACK & PLUG	ETCO	16
CONNECTOR: DIP SHORTING PLUG	Sintec	2
CONNECTOR: F- SERIES, TV, COAX	ETCO	17
CONNECTOR: F- SERIES, TV, COAX	Nemal Electronics	4

COMPONENT	COMPANY NAME	VARIETY STOCKED
CONNECTOR: F- SERIES, TV, COAX	Aldelco	19
CONNECTOR: F- SERIES, TV, COAX	Alpha Electronic Labs	5
CONNECTOR: F59, TV, COAX	Circuit Specialists	1
CONNECTOR: F59, TV, COAX	Semiconductors Surplus	1
CONNECTOR: F61, TV PANEL JACK, COAX	Circuit Specialists	1
CONNECTOR: F61, TV PANEL JACK, COAX	Semiconductors Surplus	1
CONNECTOR: F71- SERIES	Semiconductors Surplus	4
CONNECTOR: GK-12-32SL, CABLE,12 PIN	Star-Tronics	1
CONNECTOR: HEADER RECEPTACLE	Mouser Electronics	12
CONNECTOR: HEADER, DIP	See Header: DIP	
CONNECTOR: HIGH VOLTAGE, CRT	ETCO	1
CONNECTOR: HIGH VOLTAGE, SAFETY	Radiokit	2
CONNECTOR: INTERLOCK RECEPTICLE, TV	ETCO	2
CONNECTOR: INTERLOCK, TV, MALE	Aldelco	1
CONNECTOR: JONES TYPE, SOCKET, PLUG	ETCO	6
CONNECTOR: JUMPER CABLE	Quest Electronics	7
CONNECTOR: JUMPER CABLE INTRA-PROBE	Jameco Electronics	2
CONNECTOR: JUMPER, HEADER, DBL ROW	Circuit Specialists	10
CONNECTOR: M-358, "T", COAX	Radiokit	1
CONNECTOR: M-358, "T", COAX	Nemal Electronics	1
CONNECTOR: M-358, "T", COAX	Fair Radio Sales	1
CONNECTOR: M-359 ELBOW	Semiconductors Surplus	1
CONNECTOR: M-359 ELBOW	Fair Radio Sales	1
CONNECTOR: M-359 ELBOW	Nemal Electronics	1
CONNECTOR: MICROPHONE	ETCO	17
CONNECTOR: MICROPHONE	Amateur Electronic Supply	10
CONNECTOR: MICROPHONE, AMPHENOL	Fair Radio Sales	5
CONNECTOR: MICROPHONE, DIN, JACK	Mouser Electronics	7
CONNECTOR: MICROPHONE, DIN, PLUG	Mouser Electronics	5
CONNECTOR: MICROPHONE, FEMALE	Aldelco	5
CONNECTOR: MICROPHONE, MALE	Aldelco	5
CONNECTOR: MICROPHONE, XLR, JACK	Mouser Electronics	1
CONNECTOR: MICROPHONE, XLR, PLUG	Mouser Electronics	5
CONNECTOR: MODULE SOCKET & PLUG	ETCO	6
CONNECTOR: MOLEX	See Also Socket: IC, Molex	
CONNECTOR: MOLEX PIN, PC MOUNT	Jameco Electronics	1
CONNECTOR: MOLEX SET	Quest Electronics	2
CONNECTOR: MOLEX STANDARD	Daytapro Electronics	4
CONNECTOR: MOTOROLA ANTENNA PLUG	Amateur Electronic Supply	1
CONNECTOR: N- SERIES ADAPTERS	Nemal Electronics	3
CONNECTOR: N- SERIES, COAX	Alaska Microwave Lab	5
CONNECTOR: N- SERIES, COAX	Fair Radio Sales	9
CONNECTOR: N- SERIES, COAX	Semiconductors Surplus	3
CONNECTOR: N- SERIES, COAX	Nemal Electronics	8
CONNECTOR: N- SERIES, COAX	BCD Radio Parts	1
CONNECTOR: N- SERIES, COAX	Surplus Electronics	14
CONNECTOR: N- SERIES, COAX	Radiokit	1
CONNECTOR: OSM- SERIES	See Connector: SMA/OSM	
CONNECTOR: PHONE JACK, 1/4 "	BCD Radio Parts	4
CONNECTOR: PHONE JACK, 1/4"	ETCO	2
CONNECTOR: PHONE JACK, 1/4"	Circuit Specialists	4
CONNECTOR: PHONE JACK, 1/4"	Mouser Electronics	12
CONNECTOR: PHONE JACK, 1/4"	Aldelco	1
CONNECTOR: PHONE JACK, 1/8"	BCD Radio Parts	1
CONNECTOR: PHONE JACK, 2.5mm	Mouser Electronics	2
CONNECTOR: PHONE JACK, 2.5mm	Circuit Specialists	2
CONNECTOR: PHONE JACK, 2.5mm	ETCO	1
CONNECTOR: PHONE JACK, 2.5mm	Mouser Electronics	6
CONNECTOR: PHONE JACK, 3.5mm	Mouser Electronics	2

COMPONENT	COMPANY NAME	VARIETY STOCKED
CONNECTOR: PHONE JACK, 3.5mm	ETCO	1
CONNECTOR: PHONE JACK, 3.5mm	Circuit Specialists	2
CONNECTOR: PHONE JACK, 3.5mm	Mouser Electronics	8
CONNECTOR: PHONE PLUG, 1/4"	Amateur Electronic Supply	6
CONNECTOR: PHONE PLUG, 1/4"	Star-Tronics	1
CONNECTOR: PHONE PLUG, 1/4"	ETCO	4
CONNECTOR: PHONE PLUG, 1/4"	Mouser Electronics	10
CONNECTOR: PHONE PLUG, 1/4"	Circuit Specialists	3
CONNECTOR: PHONE PLUG, 1/4"	Daytapro Electronics	2
CONNECTOR: PHONE PLUG, 1/4", DUAL	Star-Tronics	1
CONNECTOR: PHONE PLUG, 1/8"	BCD Radio Parts	1
CONNECTOR: PHONE PLUG, 1/8"	Digital Research	1
CONNECTOR: PHONE PLUG, 2.5mm	BCD Radio Parts	1
CONNECTOR: PHONE PLUG, 2.5mm	Mouser Electronics	2
CONNECTOR: PHONE PLUG, 2.5mm	Circuit Specialists	1
CONNECTOR: PHONE PLUG, 3.5mm	ETCO	2
CONNECTOR: PHONE PLUG, 3.5mm	Circuit Specialists	3
CONNECTOR: PHONE PLUG, 3.5mm	Mouser Electronics	6
CONNECTOR: PHONE PLUG, 3.5mm	Nemal Electronics	1
CONNECTOR: PHONE PLUG, 3.5mm	Mouser Electronics	1
CONNECTOR: PHONE PLUG, 3/32"	BCD Radio Parts	1
CONNECTOR: PHONE PLUG, 3/32"	Digital Research	1
CONNECTOR: PHONE PLUG, 4.5mm	ETCO	1
CONNECTOR: PHONE PLUG, MINIATURE	Amateur Electronic Supply	1
CONNECTOR: PHONE PLUG, MINIATURE	Daytapro Electronics	1
CONNECTOR: PHONE PLUG, SUB-MINI	Amateur Electronic Supply	1
CONNECTOR: PHONO TYPE	Amateur Electronic Supply	3
CONNECTOR: PHONO TYPE	See Also Connector: RCA Phon	
CONNECTOR: PIN INSERT, WITH COVER	Electronic Marketplace	8
CONNECTOR: PIN JACK, 2MM, TEST PROD	ETCO	1
CONNECTOR: PIN PLUG, 2MM, TEST PROD	ETCO	1
CONNECTOR: PL- SERIES, COAX	Aldelco	1
CONNECTOR: PL- SERIES, COAX	BCD Radio Parts	2
CONNECTOR: PL- SERIES, COAX	Barker & Williamson	1
CONNECTOR: PL- SERIES, COAX	Jameco Electronics	2
CONNECTOR: PL- SERIES, COAX	Mouser Electronics	3
CONNECTOR: PL- SERIES, COAX	Alpha Electronic Labs	4
CONNECTOR: PL- SERIES, COAX	G & C Communications	2
CONNECTOR: PL- SERIES, COAX	Lacue Communications	1
CONNECTOR: PL- SERIES, COAX	Amateur Electronic Supply	4
CONNECTOR: PL- SERIES, COAX	Nemal Electronics	5
CONNECTOR: PL- SERIES, COAX	Fair Radio Sales	4
CONNECTOR: PL- SERIES, COAX	Semiconductors Surplus	2
CONNECTOR: PL- SERIES, COAX	Radiokit	2
CONNECTOR: PL- SERIES, COAX	Surplus Electronics	3
CONNECTOR: PL- SERIES, COAX	ETCO	3
CONNECTOR: POWER, .09" ID, JACK	BCD Radio Parts	1
CONNECTOR: POWER, YEASU TYPE, 2 PIN	ETCO	1
CONNECTOR: QUICK HOOK, SPRING, INS	Jameco Electronics	ASST
CONNECTOR: RCA PHONO	Amateur Electronic Supply	3
CONNECTOR: RCA PHONO JACK	BCD Radio Parts	3
CONNECTOR: RCA PHONO JACK	Circuit Specialists	4
CONNECTOR: RCA PHONO JACK	Digi-Key	1
CONNECTOR: RCA PHONO JACK	Nemal Electronics	2
CONNECTOR: RCA PHONO JACK	ETCO	5
CONNECTOR: RCA PHONO JACK TO JACK	Surplus Electronics	1
CONNECTOR: RCA PHONO JACK, DUAL	Aldelco	1
CONNECTOR: RCA PHONO JACK, DUAL	Star-Tronics	1
CONNECTOR: RCA PHONO JACK, HI-REL	Mouser Electronics	2

COMPONENT	COMPANY NAME	VARIETY STOCKED
CONNECTOR: RCA PHONO JACK, IN-LINE	Mouser Electronics	1
CONNECTOR: RCA PHONO JACK, PANEL MT	Mouser Electronics	9
CONNECTOR: RCA PHONO JACK, PANEL MT	Daytapro Electronics	1
CONNECTOR: RCA PHONO JACK, PANEL MT	Star-Tronics	1
CONNECTOR: RCA PHONO JACK, PANEL MT	Aldelco	1
CONNECTOR: RCA PHONO JACK, PC MT	Surplus Electronics	1
CONNECTOR: RCA PHONO JACK, PC MT	Mouser Electronics	1
CONNECTOR: RCA PHONO PLUG	Circuit Specialists	4
CONNECTOR: RCA PHONO PLUG	Surplus Electronics	1
CONNECTOR: RCA PHONO PLUG	Nemal Electronics	1
CONNECTOR: RCA PHONO PLUG	Daytapro Electronics	1
CONNECTOR: RCA PHONO PLUG	ETCO	5
CONNECTOR: RCA PHONO PLUG	Mouser Electronics	5
CONNECTOR: RCA PHONO PLUG	Digi-Key	3
CONNECTOR: RCA PHONO PLUG	BCD Radio Parts	1
CONNECTOR: RCA PHONO PLUG, HI-REL	Mouser Electronics	2
CONNECTOR: RIBBON CABLE	ETCO	2
CONNECTOR: RIBBON CABLE HEADER	Active Electronics	9
CONNECTOR: RIBBON CABLE, MASS TERM.	Alpha Electronic Labs	7
CONNECTOR: RIBBON CABLE, MASS TERM.	Mouser Electronics	6
CONNECTOR: RIBBON CABLE, MASS TERM.	Star-Tronics	1
CONNECTOR: RIBBON CABLE, MASS TERM.	Active Electronics	16
CONNECTOR: RIBBON CABLE-CARD EDGE	Star-Tronics	1
CONNECTOR: ROTOR CABLE PLUG & JACK	ETCO	2
CONNECTOR: RS-232	See Also Connector: D-Type	
CONNECTOR: RS-232, FEMALE	Jameco Electronics	2
CONNECTOR: RS-232, FEMALE, PC TYPE	Semiconductors Surplus	1
CONNECTOR: RS-232, MALE	Jameco Electronics	2
CONNECTOR: RS-232, MALE, 25 PIN	Star-Tronics	1
CONNECTOR: RS-232, MALE, PC TYPE	Semiconductors Surplus	1
CONNECTOR: SEALECTRO SUBMINI RF	Semiconductors Surplus	2
CONNECTOR: SMA/OSM SERIES, COAX	Alaska Microwave Lab	6
CONNECTOR: SMA/OSM SERIES, COAX	Semiconductors Surplus	17
CONNECTOR: SO- SERIES, COAX	Van Gorden Engineering	1
CONNECTOR: SO- SERIES, COAX	Surplus Electronics	2
CONNECTOR: SO- SERIES, COAX	Star-Tronics	1
CONNECTOR: SO- SERIES, COAX	Nemal Electronics	1
CONNECTOR: SO- SERIES, COAX	Daytapro Electronics	1
CONNECTOR: SO- SERIES, COAX	Semiconductors Surplus	1
CONNECTOR: SO- SERIES, COAX	Alpha Electronic Labs	1
CONNECTOR: SO- SERIES, COAX	Amp Supply	1
CONNECTOR: SO- SERIES, COAX	Amateur Electronic Supply	7
CONNECTOR: SO- SERIES, COAX	Radiokit	1
CONNECTOR: SO- SERIES, COAX	Aldelco	1
CONNECTOR: SO- SERIES, COAX	Fair Radio Sales	3
CONNECTOR: SO- SERIES, COAX	ETCO	1
CONNECTOR: SO- SERIES, COAX	Daytapro Electronics	2
CONNECTOR: SO- SERIES, COAX	Mouser Electronics	1
CONNECTOR: SO- SERIES, COAX	Jameco Electronics	1
CONNECTOR: SO-239 DIRT CAP	BCD Radio Parts	1
CONNECTOR: SPEAKER, PUSH-IN	Aldelco	2
CONNECTOR: SPEAKER, SPRING LOADED	ETCO	3
CONNECTOR: SPLICER, WIRE, STRIPLESS	Fuji-Svea	1
CONNECTOR: SPLICER, WIRE, SCOTCH LOK	Digital Research	1
CONNECTOR: STRIP LINE PLUG	Circuit Specialists	1
CONNECTOR: STRIP LINE PLUG	Mouser Electronics	4
CONNECTOR: SWITCHBOARD JACK, 1/4"	Atlantic Surplus Sales	1
CONNECTOR: TELEPHONE JACK	Daytapro Electronics	3
CONNECTOR: TELEPHONE JACK	ETCO	1

COMPONENT	COMPANY NAME	VARIETY STOCKED
CONNECTOR: TELEPHONE JACK, MINI	Daytapro Electronics	1
CONNECTOR: TELEPHONE JACK/PLUG	ETCO	6
CONNECTOR: TELEPHONE MODULAR ADAPTR	ETCO	4
CONNECTOR: TELEPHONE MODULAR JACK	ETCO	1
CONNECTOR: TELEPHONE MODULAR PLUG	ETCO	1
CONNECTOR: TELEPHONE PLUG ADAPTER	ETCO	2
CONNECTOR: TEST PIN, STANDARD	Mouser Electronics	5
CONNECTOR: TEST PROBE	ETCO	2
CONNECTOR: TEST PROBE, INSULATED	Digi-Key	2
CONNECTOR: TEST PROBE, SEIZER	ETCO	1
CONNECTOR: TEST PROD, STANDARD	Mouser Electronics	6
CONNECTOR: TIP JACK	ETCO	1
CONNECTOR: TIP JACK, HORIZONTAL, PC	Digi-Key	24
CONNECTOR: TIP JACK, PANEL MOUNT	Digi-Key	10
CONNECTOR: TIP JACK, PC MOUNT	Star-Tronics	1
CONNECTOR: TIP JACK, VERTICAL	Digi-Key	5
CONNECTOR: TIP PLUG	ETCO	1
CONNECTOR: TIP PLUG	Mouser Electronics	25
CONNECTOR: TIP PLUG, SOLDERLESS	Digi-Key	2
CONNECTOR: TWIN LEAD, TV, 300 OHM	ETCO	1
CONNECTOR: UG- SERIES REDUCERS	Radiokit	2
CONNECTOR: UG- SERIES REDUCERS	Lacue Communications	2
CONNECTOR: UG- SERIES REDUCERS	G & C Communications	2
CONNECTOR: UG- SERIES REDUCERS	Jameco Electronics	1
CONNECTOR: UG- SERIES REDUCERS	Aldelco	2
CONNECTOR: UG- SERIES REDUCERS	Amateur Electronic Supply	2
CONNECTOR: UG- SERIES REDUCERS	Fair Radio Sales	2
CONNECTOR: UG- SERIES REDUCERS	Surplus Electronics	2
CONNECTOR: UG- SERIES REDUCERS	Nemal Electronics	2
CONNECTOR: UG- SERIES REDUCERS	BCD Radio Parts	1
CONNECTOR: UG- SERIES REDUCERS	ETCO	2
CONNECTOR: UG- SERIES, ADAPTERS	Alpha Electronic Labs	6
CONNECTOR: UG-100,109,160,212 COAX	Fair Radio Sales	4
CONNECTOR: UG-363, DBLE F, PANEL MT	Semiconductors Surplus	1
CONNECTOR: UHF "T"	Alpha Electronic Labs	1
CONNECTOR: UHF ADAPTERS	Aldelco	8
CONNECTOR: UHF PLUG	Radiokit	1
CONNECTOR: UHF RIGHT ANGLE BEND	Star-Tronics	1
CONNECTOR: UHF SERIES	See Also Connector: SO-	
CONNECTOR: UHF SERIES	See Also Connector: PL-	
CONNECTOR: UHF SERIES, ADAPTERS	ETCO	12
CONNECTOR: UHF SERIES, ADAPTERS	Surplus Electronics	4
CONNECTOR: VIDEO PLUG, JACK, 10 PIN	ETCO	2
CONNECTOR: VIDEO PLUG, JACK, 8 PIN	ETCO	2
CONNECTOR: VTR, INLINE JACK, 10 PIN	ETCO	1
CONNECTOR: XLR, HI-REL	Mouser Electronics	4
CONVERTER: TVC-2, 435 MHZ TO TV	P.C. Electronics	1
CORD: ABRASIVE	Small Parts	23
CORD: AC	See Cable: AC Cord	
CORD: KEVLAR FIBER	Edmund Scientific	1
CORE: ANTENNA BALUN	Semiconductors Surplus	4
CORE: ANTENNA BALUN, FERRITE, KIT	Amidon Associates	1
CORE: ANTENNA BALUN, IRON CORE, KIT	Amidon Associates	1
CORE: POT, FERRRITE, AUDIO-2 MHZ	Radiokit	1
CORE: TC902, CUP, FERRITE, 11/16"OD	Star-Tronics	1
CORE: TOROIDAL	Digital Research	1

COMPONENT	COMPANY NAME	VARIETY STOCKED
CORE: TOROIDAL	Circuit Specialists	9
CORE: TOROIDAL, FERRITE	Amidon Associates	51
CORE: TOROIDAL, FERRITE	Star-Tronics	1
CORE: TOROIDAL, FERRITE	Palomar Engineers	13
CORE: TOROIDAL, FERRITE	Radiokit	20
CORE: TOROIDAL, FERRITE, HIGH POWER	Amidon Associates	10
CORE: TOROIDAL, FERRITE, JW MILLER	Mouser Electronics	12
CORE: TOROIDAL, IRON PDR, JW MILLER	Mouser Electronics	13
CORE: TOROIDAL, IRON POWDER	Semiconductors Surplus	11
CORE: TOROIDAL, IRON POWDER	Radiokit	41
CORE: TOROIDAL, IRON POWDER	Amidon Associates	116
CORE: TOROIDAL, IRON POWDER	ETCO	18
CORE: TOROIDAL, IRON POWDER	Palomar Engineers	16
CORE: TOROIDAL, WIDE BAND XFMR	Circuit Specialists	2
CORE: WBT65-, WB XFMR, JW MILLER	Mouser Electronics	4
CORK: SHEET, COMPRESSIBLE	Small Parts	2
COUNTER: MECHANICAL	Fair Radio Sales	1
COUNTER: MECHANICAL, 24 VDC COIL	Star-Tronics	1
COUNTER: MECHANICAL, 60 VDC	ETCO	1
COUNTER: MECHANICAL, DIGITAL, 24VDC	BCD Radio Parts	1
COUPLER: 1.5-125 MHZ, 2 PORT	Elcom Systems	2
COUPLER: 1.5-125 MHZ, 4 PORT	Elcom Systems	2
COUPLER: CROSS GUIDE, WAVEGUIDE	Lectronic	
COUPLER: DIAL SHAFT	See Dial: Coupler	
COUPLER: DIRECTIONAL, MICROWAVE	Lectronic	
COUPLER: DIRECTIONAL, TO 2 GHZ	Mini-Circuits Laboratory	35
COUPLER: MICROWAVE, CAPACITIVE	Elcom Systems	1
CRT:	See Tube: Cathode Ray	
CRYSTAL DISCRIMINATOR: 9 & 10.7 MHZ	Spectrum International	4
CRYSTAL SOCKET:	See Socket	
CRYSTAL:	Certified International	
CRYSTAL:	See Also Oscillator: Crystal	
CRYSTAL: 100 KHZ IN OVEN, 115V	Fair Radio Sales	1
CRYSTAL: 100 KHZ, HIGH TOLERANCE	ETCO	1
CRYSTAL: 3.58 MHZ, COLOR BURST	Fuji-Svea	1
CRYSTAL: 32,768 HZ, MICRO, WATCH	Semiconductors Surplus	1
CRYSTAL: 32,768 HZ, MICRO, WATCH	Star-Tronics	1
CRYSTAL: 3579.545 KHZ PRECISION	ETCO	1
CRYSTAL: 5.24288 MHZ	Circuit Specialists	1
CRYSTAL: ADAPTER	International Crystal	2
CRYSTAL: AIRCRAFT BAND	Rolin Distributors	
CRYSTAL: AMATEUR 2 METER BAND	Rolin Distributors	
CRYSTAL: AMATEUR BAND	Savoy Electronics	
CRYSTAL: AMATEUR BAND	C-W Crystals	
CRYSTAL: BC EQUIPEMENT	Fair Radio Sales	
CRYSTAL: C.B. WALKIE TALKIE	Rolin Distributors	
CRYSTAL: C.B. XMIT, RCVR	Semiconductors Surplus	80
CRYSTAL: C.B. XMIT, RCVR, SYNTH	Rolin Distributors	
CRYSTAL: CARRIER, 9 & 10.7 MHZ IF	Spectrum International	4
CRYSTAL: GENERAL PURPOSE	C-W Crystals	
CRYSTAL: HF, 30 MHZ- 160 MHZ	International Crystal	
CRYSTAL: HF, 30 MHZ- 200 MHZ	Spectrum International	

COMPONENT	COMPANY NAME	VARIETY STOCKED
CRYSTAL: HF, 30 MHZ- 60 MHZ	Rolin Distributors	
CRYSTAL: LAND MOBILE, COMMERCIAL	Sentry Manufacturing	
CRYSTAL: LF, 1 KHZ- 999 KHZ	Spectrum International	
CRYSTAL: LF, 5 KHz- 999 KHz	Rolin Distributors	
CRYSTAL: LF, 70 KHZ- 999 KHZ	International Crystal	
CRYSTAL: MARINE BAND, VHF	Rolin Distributors	
CRYSTAL: MF, 1 MHZ- 30 MHZ	Spectrum International	
CRYSTAL: MF, 1 MHZ- 30 MHZ	International Crystal	
CRYSTAL: MF, 1 MHZ- 30 MHZ	Rolin Distributors	
CRYSTAL: MICROPROCESSOR SYSTEM	Jameco Electronics	34
CRYSTAL: MICROPROCESSOR SYSTEM	Mouser Electronics	30
CRYSTAL: MICROPROCESSOR SYSTEM	Piezo Technology	
CRYSTAL: MICROPROCESSOR SYSTEM	Sentry Manufacturing	
CRYSTAL: MICROPROCESSOR SYSTEM	Electronic Marketplace	
CRYSTAL: MONITOR SCANNER	Sentry Manufacturing	
CRYSTAL: MONITOR SCANNER	Rolin Distributors	
CRYSTAL: PRE-CUT	Active Electronics	29
CRYSTAL: PRE-CUT	Circuit Specialists	11
CRYSTAL: PRE-CUT	Digital Research	10
CRYSTAL: PRE-CUT	BCD Radio Parts	22
CRYSTAL: PRE-CUT	Quest Electronics	39
CRYSTAL: PRE-CUT	B.G. Micro	13
CRYSTAL: PRE-CUT, 2- 6.5536 MHZ	Digi-Key	35
CRYSTAL: PRE-CUT, 5.12- 60 MHZ	Semiconductors Surplus	80
DC/DC CONVERTER: 2 V IN, -25V OUT	ETCO	1
DC/DC CONVERTER: 5 VDC IN, LV OUT	Sintec	12
DC/DC CONVERTER: LV IN, LV OUT	Active Electronics	18
DECOUPLER: MICROWAVE, INDUCTIVE	Elcom Systems	1
DECOUPLER: MICROWAVE, RESISTIVE	Elcom Systems	1
DELAY LINE: AUDIO, DUAL, VIBRASONIC	Babylon Electronics	1
DETECTOR: COAXIAL, 50 OHM,.01-4 GHZ	Elcom Systems	1
DETECTOR: COAXIAL, CRYSTAL, MICROWV	Lectronic	
DIAC:	See Transistor: DIAC	
DIAL: 31227-3 HAMMARLUND	Fair Radio Sales	1
DIAL: BEARING, PANEL	Radiokit	1
DIAL: CORD, NYLON	ETCO	3
DIAL: COUNTER AND SCALE, MILLEN	Radiokit	1
DIAL: COUNTING, FOR 10 TURN POT	ETCO	1
DIAL: COUPLER, BRASS, 1/4"	ETCO	2
DIAL: COUPLER, CERAMIC	Fair Radio Sales	4
DIAL: COUPLER, FLEXIBLE, 1/4"	ETCO	2
DIAL: COUPLER, FLEXIBLE, MILLEN	Radiokit	11
DIAL: COUPLER, POLYURETHANE	Small Parts	14
DIAL: COUPLER, STEATITE, 6mm SHAFT	Fuji-Svea	1
DIAL: CYCLOMETER-TYPE COUNTER	Barker & Williamson	1
DIAL: CYCLOMETER-TYPE COUNTER, B&W	Radiokit	1
DIAL: CYCLOMETER-TYPE COUNTER, B&W	Amp Supply	1
DIAL: DRIVE	See Also Drive:	
DIAL: DRIVE UNIT, MILLEN	Radiokit	1
DIAL: DRIVE, 10:1 EPICYCLIC	Radiokit	1
DIAL: DRIVE, 6:1 REDUCTION RATIO	Radiokit	1
DIAL: DRIVE, DUAL RATIO	Radiokit	1
DIAL: DRIVE, DUAL RATIO BALL, SCALE	Radiokit	1

COMPONENT	COMPANY NAME	VARIETY STOCKED
DIAL: DRIVE, PANEL MOUNT	ETCO	4
DIAL: DRIVE, PLANETARY DRIVE	Amp Supply	1
DIAL: DRIVE, PLANETARY DRIVE	ETCO	1
DIAL: DRIVE, RIGHT ANGLE	Fair Radio Sales	1
DIAL: DRIVE, RIGHT ANGLE, MILLEN	Radiokit	1
DIAL: FOR 10-TURN VARIABLE RESISTOR	Circuit Specialists	1
DIAL: LARGE BALL DRIVE, SCALE	Radiokit	1
DIAL: LOCK	Radiokit	5
DIAL: SHAFT	See Also Shaft:	
DIAL: SHAFT EXTENDER, 1/4" x 9"	ETCO	1
DIAL: SHAFT EXTENDER, FLEXIBLE, 8"	ETCO	1
DIAL: SHAFT LOCK	ETCO	3
DIAL: SHAFT REDUCER, 1/4 TO 1/8"	Radiokit	1
DIAL: VERNIER DRIVE, ENGRAVED	Mouser Electronics	2
DIAL: VERNIER, 0-100 SCALE, 180 DEG	Fuji-Svea	1
DIAL: VERNIER, 0-100 SCALE, 300 DEG	Fuji-Svea	1
DIAL: VERNIER, 8:1 DRIVE RATIO	ETCO	3
DIAL: VERNIER, 8:1 DRIVE RATIO	Radiokit	1
DIAL: VERNIER, HAIRLINE POINTER	ETCO	1
DIODE: 1 KV PIV, 3 AMP	Amp Supply	1
DIODE: 10 KV PIV, 10mA	Star-Tronics	1
DIODE: 10 KV PIV, 15mA	Semiconductors Surplus	1
DIODE: 100 PIV, 15 AMP	Star-Tronics	1
DIODE: 1000 PIV, 2.5A	Semiconductors Surplus	1
DIODE: 15A, STUD MOUNT	Circuit Specialists	5
DIODE: 20 KV PIV, 25 mA	Semiconductors Surplus	1
DIODE: 200 PIV, 100 AMP	Fair Radio Sales	1
DIODE: 200 PIV, 6 AMP	Semiconductors Surplus	1
DIODE: 400 PIV, 35 AMP	Star-Tronics	1
DIODE: 5 KV PIV, 250 mA	Digital Research	1
DIODE: 5 KV PIV, 50mA	Semiconductors Surplus	1
DIODE: 10D- SERIES, NIHON INTER	Fuji-Svea	2
DIODE: 1N- SERIES	Aldelco	132
DIODE: 1N- SERIES	Quest Electronics	
DIODE: 1N- SERIES	Proverty Electronics	4
DIODE: 1N- SERIES	Digital Research	3
DIODE: 1N- SERIES, SWITCHING	Active Electronics	19
DIODE: 1N1100 SERIES, RECTIFIER,35A	Alpha Electronic Labs	4
DIODE: 1N1100 SERIES, RECTIFIER,35A	Jameco Electronics	6
DIODE: 1N1200 SERIES, RECTIFIER,12A	Jameco Electronics	4
DIODE: 1N23DR MICROWAVE	ETCO	1
DIODE: 1N270, SIGNAL	Alpha Electronic Labs	1
DIODE: 1N34, GERMANIUM, SIGNAL	Active Electronics	1
DIODE: 1N34, GERMANIUM, SIGNAL	Mouser Electronics	1
DIODE: 1N34, GERMANIUM, SIGNAL	Alpha Electronic Labs	1
DIODE: 1N34, GERMANIUM, SIGNAL	Radiokit	1
DIODE: 1N34, GERMANIUM, SIGNAL	Fuji-Svea	1
DIODE: 1N4000 SERIES, RECTIFIER	Radiokit	3
DIODE: 1N4000 SERIES, RECTIFIER	Active Electronics	7
DIODE: 1N4000 SERIES, RECTIFIER	Electronic Marketplace	2
DIODE: 1N4000 SERIES, RECTIFIER	Digi-Key	7
DIODE: 1N4000 SERIES, RECTIFIER	Circuit Specialists	5
DIODE: 1N4000 SERIES, RECTIFIER	Digital Research	1
DIODE: 1N4000 SERIES, RECTIFIER	Jameco Electronics	7
DIODE: 1N4000 SERIES, RECTIFIER	Mouser Electronics	7
DIODE: 1N4000 SERIES, RECTIFIER	Alpha Electronic Labs	7
DIODE: 1N4000 SERIES, RECTIFIER	Quest Electronics	7
DIODE: 1N4000 SERIES, RECTIFIER	Fuji-Svea	7

COMPONENT	COMPANY NAME	VARIETY STOCKED
DIODE: 1N4005 SERIES, RECTIFIER	Key Electronics	1
DIODE: 1N4148, SILICON, SIGNAL	Star-Tronics	1
DIODE: 1N4148, SILICON, SIGNAL	Digital Research	1
DIODE: 1N4148, SILICON, SIGNAL	Fuji-Svea	1
DIODE: 1N4148, SILICON, SIGNAL	Jameco Electronics	1
DIODE: 1N4148, SILICON, SIGNAL	Proverty Electronics	1
DIODE: 1N4148, SILICON, SIGNAL	Electronic Marketplace	1
DIODE: 1N4148, SILICON, SIGNAL	Digi-Key	1
DIODE: 1N4148, SILICON, SIGNAL	Mouser Electronics	1
DIODE: 1N4148, SILICON, SIGNAL	Quest Electronics	1
DIODE: 1N4700 SERIES, RECTIFIER, 3A	Quest Electronics	7
DIODE: 1N5000 SERIES, RECTIFIER, 3A	Alpha Electronic Labs	4
DIODE: 1N5000 SERIES, RECTIFIER, 3A	Jameco Electronics	6
DIODE: 1N5000 SERIES, RECTIFIER, 3A	Active Electronics	8
DIODE: 1N5000 SERIES, RECTIFIER, 3A	Mouser Electronics	6
DIODE: 1N5148, TUNING DIODE	Fuji-Svea	1
DIODE: 1N5711, MICROWAVE MIXER	MHZ Electronics	1
DIODE: 1N60, GERMANIUM, SIGNAL	Mouser Electronics	1
DIODE: 1N60, GERMANIUM, SIGNAL	Fuji-Svea	1
DIODE: 1N6263, SCHOTTKY	Mouser Electronics	1
DIODE: 1N914, ARRAY OF 10	Babylon Electronics	1
DIODE: 1N914, SWITCHING	Quest Electronics	1
DIODE: 1N914, SWITCHING	Babylon Electronics	1
DIODE: 1N914, SWITCHING	Electronic Marketplace	1
DIODE: 1N914, SWITCHING	Semiconductors Surplus	1
DIODE: 1N914, SWITCHING	Radiokit	1
DIODE: 1N914, SWITCHING	Alpha Electronic Labs	1
DIODE: 1N914, SWITCHING	Key Electronics	1
DIODE: 1N914, SWITCHING	Circuit Specialists	1
DIODE: 1S- SERIES, JAPANESE	Fuji-Svea	13
DIODE: 1S- SERIES, TUNNEL	MHZ Electronics	2
DIODE: 1SS 97 UHF MIXER	Active Electronics	1
DIODE: 1SS 98 MICROWAVE SCHOTTKY	BCD Radio Parts	1
DIODE: 1SS 98 UHF DETECTOR	Active Electronics	1
DIODE: 221- SERIES, ZENITH	Fuji-Svea	16
DIODE: 2XXXX SERIES, TUNING	Circuit Specialists	6
DIODE: 3N- SERIES, BRIDGE RECTIFIER	Circuit Specialists	4
DIODE: 5082-2835 MICROWAVE	Digital Research	1
DIODE: 5082-2835 MICROWAVE, HP	BCD Radio Parts	1
DIODE: BRIDGE	Electronic Marketplace	2
DIODE: BRIDGE, FULL WAVE, 1-35 A	Active Electronics	52
DIODE: BRIDGE, FULL WAVE, 1.5-25 A	Mouser Electronics	4
DIODE: BRIDGE, FULL WAVE, 6-25 A	Star-Tronics	2
DIODE: BRIDGE, FULL WAVE, 6-25 AMP	BCD Radio Parts	3
DIODE: BRIDGE, MINI DIP CASE, 1A	BCD Radio Parts	2
DIODE: CATSWHISKER CRYSTAL DETECTOR	ETCO	2
DIODE: CURRENT REGULATOR	Fuji-Svea	4
DIODE: DM- SERIES, DELCO	Fuji-Svea	11
DIODE: EK500, 5 KV PIV, 250mA	Digital Research	1
DIODE: FD- SERIES, 7 nS	Active Electronics	2
DIODE: FD- SERIES, LOW LEAKAGE	Active Electronics	4
DIODE: FDH- SERIES, 4 nS	Active Electronics	2
DIODE: FDH- SERIES, LOW LEAKAGE	Active Electronics	2
DIODE: FDH- SERIES, SWITCHING	Active Electronics	2
DIODE: FJT 1100 PICO AMP LEAKAGE	Active Electronics	1
DIODE: GERMANIUM	ETCO	1
DIODE: GERMANIUM	Mouser Electronics	12
DIODE: HN-1, 4 GHZ, HOT CARRIER	Alaska Microwave Lab	1
DIODE: LED	See LED:	

COMPONENT	COMPANY NAME	VARIETY STOCKED
DIODE: MA41482, 10-40 GHZ	Semiconductors Surplus	1
DIODE: MBD 101 & 102, HOT CARRIER	Fuji-Svea	2
DIODE: MBD 101, UHF, HOT CARRIER	Alaska Microwave Lab	3
DIODE: MBD 101, UHF, HOT CARRIER	Alpha Electronic Labs	1
DIODE: MBD 101, UHF, HOT CARRIER	Circuit Specialists	1
DIODE: MDA- SERIES BRIDGE RECTIFIER	Jameco Electronics	3
DIODE: MDA- SERIES BRIDGE RECTIFIER	Circuit Specialists	12
DIODE: MDA- SERIES BRIDGE RECTIFIER	Fuji-Svea	39
DIODE: MPN 3401, PIN RF SWITCHING	Circuit Specialists	1
DIODE: MPN 3401, PIN RF SWITCHING	Fuji-Svea	1
DIODE: MPN 3401, PIN RF SWITCHING	Alpha Electronic Labs	1
DIODE: MPN 3401, PIN RF SWITCHING	Semiconductors Surplus	1
DIODE: MPN 3402, PIN RF SWITCHING	Alpha Electronic Labs	1
DIODE: MR- SERIES RECTIFIER	Alpha Electronic Labs	18
DIODE: MR- SERIES RECTIFIER	Fair Radio Sales	4
DIODE: MR- SERIES RECTIFIER	Fuji-Svea	35
DIODE: MR- SERIES RECTIFIER	Circuit Specialists	12
DIODE: MR- SERIES RECTIFIER, 6A	Active Electronics	7
DIODE: MR1- SERIES, RECT, DAMPER	Fuji-Svea	2
DIODE: MV- SERIES, TUNING	Fuji-Svea	5
DIODE: MV- SERIES, TUNING VARACTER	Alpha Electronic Labs	8
DIODE: MV830-MV840, TUNING	Circuit Specialists	11
DIODE: MZ2201 SURMETIC REFERENCE	Fuji-Svea	1
DIODE: ND4000 SERIES, SCHOTTKY	California Eastern Lab	8
DIODE: ND4131, 4 GHZ, HOT CARRIER	Alaska Microwave Lab	1
DIODE: ND487 SERIES, SCHOTTKY, QUAD	California Eastern Lab	8
DIODE: ND4900 SERIES, BATCH MATCHED	California Eastern Lab	4
DIODE: ND4900 SERIES, SCHOTTKY BARR	California Eastern Lab	4
DIODE: ND5 SERIES, GaAs SCHOTTKY	California Eastern Lab	4
DIODE: ND6 SERIES, PIN	California Eastern Lab	11
DIODE: ND7 SERIES, GUNN	California Eastern Lab	14
DIODE: ND8 SERIES, CW IMPATT	California Eastern Lab	14
DIODE: NDL SERIES, AVALANCHE	California Eastern Lab	3
DIODE: NDL SERIES, LASER	California Eastern Lab	7
DIODE: NDL SERIES, PIN	California Eastern Lab	6
DIODE: PR 5400, RECTIFIER	Digi-Key	1
DIODE: RECTIFIER	See Also Specific Part #	
DIODE: RECTIFIER	Mouser Electronics	51
DIODE: RECTIFIER	Fair Radio Sales	11
DIODE: RECTIFIER	Fuji-Svea	51
DIODE: RECTIFIER	Aldelco	21
DIODE: RECTIFIER, BRIDGE	Babylon Electronics	2
DIODE: RECTIFIER, BRIDGE	Digital Research	6
DIODE: RECTIFIER, BRIDGE	Aldelco	12
DIODE: RECTIFIER, BRIDGE	Semiconductors Surplus	10
DIODE: RECTIFIER, FAST RECOVERY	Fuji-Svea	9
DIODE: RECTIFIER, POWER	ETCO	19
DIODE: RECTIFIER, SCHOTTKY	Fuji-Svea	3
DIODE: RECTIFIER, SCHOTTKY	Mouser Electronics	4
DIODE: S1-1, RECTIFIER	Digital Research	1
DIODE: SCR	See Transistor: SCR	
DIODE: SD- SERIES, SCHOTTKY, POWER	Active Electronics	2
DIODE: SELENIUM RECTIFIER	ETCO	7
DIODE: SILICON, SIGNAL	Mouser Electronics	17
DIODE: TRANSIENT VOLTAGE SUPPRESSOR	Active Electronics	14
DIODE: TRANSIENT VOLTAGE SUPPRESSOR	Mouser Electronics	6
DIODE: TRANSIENT VOLTAGE SUPPRESSOR	BCD Radio Parts	1
DIODE: TRANSIENT VOLTAGE SUPPRESSOR	Digital Research	1
DIODE: TRIAC	See Transistor: TRIAC	

COMPONENT	COMPANY NAME	VARIETY STOCKED
DIODE: TUNING, MOTOROLA	Mouser Electronics	25
DIODE: UES 1302 HI EFFICIENCY RECT	Active Electronics	1
DIODE: VARICAP, CRT	Mouser Electronics	3
DIODE: VIDEO DETECTOR,'9 GHZ TEST	Lectronic	1
DIODE: VO- SERIES, HITACHI	Fuji-Svea	2
DIODE: ZENER REFERENCE	Babylon Electronics	2
DIODE: ZENER REFERENCE, .02-1 %	Fuji-Svea	5
DIODE: ZENER, 1 WATT	Jameco Electronics	7
DIODE: ZENER, 1 WATT	Alpha Electronic Labs	
DIODE: ZENER, 1 WATT	Active Electronics	
DIODE: ZENER, 1 WATT	Aldelco	18
DIODE: ZENER, 1 WATT	Electronic Marketplace	2
DIODE: ZENER, 1 WATT	Fuji-Svea	83
DIODE: ZENER, 1 WATT	Mouser Electronics	28
DIODE: ZENER, 1 WATT	Key Electronics	6
DIODE: ZENER, 1 WATT	Digital Research	4
DIODE: ZENER, 1 WATT	Fair Radio Sales	6
DIODE: ZENER, 1 WATT	Circuit Specialists	31
DIODE: ZENER, 1 WATT, 75 VOLT	Star-Tronics	1
DIODE: ZENER, 1 WATT, MOTOROLA	Mouser Electronics	29
DIODE: ZENER, 10 WATT	Fuji-Svea	2
DIODE: ZENER, 10 WATT	Fair Radio Sales	2
DIODE: ZENER, 10 WATT	Digital Research	1
DIODE: ZENER, 10 WATT	Mouser Electronics	6
DIODE: ZENER, 3 WATT	Active Electronics	2
DIODE: ZENER, 3 WATT	Electronic Marketplace	1
DIODE: ZENER, 400 mW	Aldelco	22
DIODE: ZENER, 400 mW	Circuit Specialists	9
DIODE: ZENER, 400 mW	Mouser Electronics	32
DIODE: ZENER, 400 mW	Star-Tronics	5
DIODE: ZENER, 400 mW	Jameco Electronics	9
DIODE: ZENER, 400 mW	Active Electronics	
DIODE: ZENER, 400 mW	Fuji-Svea	7
DIODE: ZENER, 400 mW	Digital Research	1
DIODE: ZENER, 5 WATT	Circuit Specialists	33
DIODE: ZENER, 5 WATT	Mouser Electronics	16
DIODE: ZENER, 5 WATT	Digital Research	9
DIODE: ZENER, 5 WATT	Active Electronics	
DIODE: ZENER, 5 WATT	Fuji-Svea	26
DIODE: ZENER, 5 WATT	Alpha Electronic Labs	
DIODE: ZENER, 5 WATT	Aldelco	20
DIODE: ZENER, 50 WATT, 8.2 V	Amp Supply	1
DIODE: ZENER, 500 mW	Jameco Electronics	6
DIODE: ZENER, 500 mW	Digi-Key	33
DIODE: ZENER, 500 mW	Electronic Marketplace	6
DIODE: ZENER, 500 mW	Active Electronics	
DIODE: ZENER, 500 mW	Mouser Electronics	43
DIODE: ZENER, 500 mW	Semiconductors Surplus	23
DIODE: ZENER, 500 mW	Fuji-Svea	127
DIODE: ZENER, 500 mW	Alpha Electronic Labs	
DIP HEADER:	See Header: DIP	
DIP PLUG:	See Also: Header	
DIP PLUG: RIBBON CABLE, WITH COVER	Digi-Key	3
DIRECTIONAL COUPLER:	See Coupler: Directional	
DISC: BRASS	Small Parts	18

COMPONENT	COMPANY NAME	VARIETY STOCKED
DISCRIMINATOR: CRYSTAL, 10.7 MHZ	Piezo Technology	1
DISCRIMINATOR: CRYSTAL, 21.4 MHZ	Piezo Technology	1
DISPLAY:	See Also LED:	
DISPLAY: 1352-1382, 7-SEGMENT	Circuit Specialists	7
DISPLAY: 1704, 5X7 ALPHANUMERIC	Mouser Electronics	1
DISPLAY: 1704, 5X7 ALPHANUMERIC	Circuit Specialists	1
DISPLAY: 1720, 1723, 7-SEGMENT	Circuit Specialists	2
DISPLAY: 1737, 1738, 7-SEGMENT	Circuit Specialists	2
DISPLAY: 1800-1825, 7-SEGMENT	Circuit Specialists	6
DISPLAY: 4520A, 4 DIGIT, 1/2"	Semiconductors Surplus	1
DISPLAY: 5082- SERIES, LED, HEW PA	BCD Radio Parts	6
DISPLAY: 5082- SERIES, LED, HEW PA	Jameco Electronics	21
DISPLAY: 5082- SERIES, LED, HEW PA	Digital Research	4
DISPLAY: 5082- SERIES, LED, HEW PA	Alpha Electronic Labs	2
DISPLAY: 5082-4697, LED, 7-SEG, HP	Fair Radio Sales	1
DISPLAY: B646(5R), 8 LED PACKAGE	Semiconductors Surplus	1
DISPLAY: BAR GRAPH, 10-SEGMENT	Jameco Electronics	6
DISPLAY: CLOCK, LED	Jameco Electronics	2
DISPLAY: DG-12, FLUORESCENT	Quest Electronics	1
DISPLAY: DL- SERIES, LED	Active Electronics	6
DISPLAY: DL- SERIES, LED	Jameco Electronics	15
DISPLAY: DL- SERIES, LED	Aldelco	4
DISPLAY: DL- SERIES, LED, 7-SEGMENT	Quest Electronics	8
DISPLAY: DL- SERIES, LED, 7-SEGMENT	Electronic Marketplace	2
DISPLAY: DL-4509, LITRONIX, 4 DIGIT	Semiconductors Surplus	1
DISPLAY: FLUORESCENT, WITH DRIVERS	Fair Radio Sales	3
DISPLAY: FND 357, LED, 7-SEGMENT	Surplus Electronics	1
DISPLAY: FND 810, LED, 7-SEG, 0.8"	BCD Radio Parts	1
DISPLAY: FND 847, LED, 7-SEG, 0.8"	Digital Research	1
DISPLAY: FND- SERIES, LED	Jameco Electronics	10
DISPLAY: FND- SERIES, LED	Digital Research	2
DISPLAY: FND- SERIES, LED	Active Electronics	9
DISPLAY: FND- SERIES, LED, 7-SEG	Babylon Electronics	2
DISPLAY: FND- SERIES, LED, 7-SEG	Aldelco	9
DISPLAY: FND- SERIES, LED, 7-SEG	Quest Electronics	5
DISPLAY: FND- SERIES, LED, 7-SEG	BCD Radio Parts	2
DISPLAY: GAS DISCHARGE, 15-SEGMENT	ETCO	1
DISPLAY: GAS DISCHARGE, 7-SEGMENT	ETCO	4
DISPLAY: GAS DISCHARGE, NIXIE	ETCO	1
DISPLAY: HDSP- SERIES, LED, HEW PA	Jameco Electronics	3
DISPLAY: INCANDESCENT, 7-SEGMENT	Mouser Electronics	2
DISPLAY: INCANDESCENT, 7-SEGMENT	ETCO	1
DISPLAY: INCANDESCENT, TUBE, 7-SEG	Mouser Electronics	2
DISPLAY: LED, 10 BAR DISPLAY	Semiconductors Surplus	1
DISPLAY: LED, 4 DIGIT, CLOCK FORMAT	Mouser Electronics	2
DISPLAY: LED, 7-SEG, DIRECT DRIVE	Digi-Key	20
DISPLAY: LED, 7-SEGMENT	Key Electronics	2
DISPLAY: LED, 7-SEGMENT	ETCO	11
DISPLAY: LED, 7-SEGMENT	Mouser Electronics	28
DISPLAY: LED, 7-SEGMENT WITH LOGIC	Circuit Specialists	3
DISPLAY: LED, 7-SEGMENT WITH LOGIC	Mouser Electronics	3
DISPLAY: LED, 7-SEGMENT, BOWMAR	BCD Radio Parts	3
DISPLAY: LED, 7-SEGMENT, BOWMAR	Digital Research	2
DISPLAY: LED, 7-SEGMENT, OPCOA	Sintec	4
DISPLAY: LED, 7-SEGMENT, WEST. ELEC	Digital Research	1
DISPLAY: LED, DOT MATRIX, 5X5	Aldelco	1
DISPLAY: LED, DUAL COLON	Semiconductors Surplus	1

COMPONENT	COMPANY NAME	VARIETY STOCKED
DISPLAY: LIQUID CRYSTAL	Circuit Specialists	6
DISPLAY: LIQUID CRYSTAL, 4-DIGIT	Jameco Electronics	1
DISPLAY: LIQUID CRYSTAL, FAIRCHILD	Sintec	4
DISPLAY: LIQUID CRYSTAL,3 1/2 DIGIT	Semiconductors Surplus	1
DISPLAY: MAN- SERIES, LED	Jameco Electronics	24
DISPLAY: MAN- SERIES, LED, 7-SEG	Quest Electronics	10
DISPLAY: MAN- SERIES, LED, 7-SEG	Aldelco	1
DISPLAY: MAN- SERIES, LED, 7-SEG	Active Electronics	13
DISPLAY: MAN- SERIES, LED, 7-SEG	Electronic Marketplace	5
DISPLAY: MAN- SERIES, LED, 7-SEG	Alpha Electronic Labs	3
DISPLAY: NSB- SERIES, LED, 7-SEG	Digi-Key	9
DISPLAY: NSM3914, 15 BAR DISPLAY	Circuit Specialists	2
DISPLAY: NSM3914, 15, 16 BAR DISPLY	Digi-Key	3
DISPLAY: NSN 381, LED, 7-SEG, DUAL	Digital Research	1
DISPLAY: NSN 381, LED, MULTIPLEXED	BCD Radio Parts	1
DISPLAY: NSN- SERIES, LED, 7-SEG	Jameco Electronics	1
DISPLAY: NSN- SERIES, LED, 7-SEG	Digi-Key	14
DISPLAY: PLASMA, 9 DIGIT, SPERRY	Babylon Electronics	1
DISPLAY: SLA-1, LED, 7-SEGMENT	BCD Radio Parts	1
DISPLAY: SLA-1, LED, 7-SEGMENT	Digital Research	1
DISPLAY: SLA-2, LED, 7-SEGMENT	Fair Radio Sales	3
DISPLAY: SP- SERIES, GAS DISCHARGE	BCD Radio Parts	2
DISPLAY: SP- SERIES, GAS DISCHARGE	Digital Research	2
DISPLAY: SUPER BRIGHT, LED, 7-SEG	Active Electronics	16
DISPLAY: TIL- SERIES, LED	Jameco Electronics	2
DISPLAY: TIL- SERIES, LED	Active Electronics	6
DISPLAY: TIL305, LED, ALPHA/NUMERIC	Quest Electronics	1
DISPLAY: TIL305P, 5X7 ALPHA/NUMERIC	Semiconductors Surplus	1
DISPLAY: TIL311, LED, HEXADECIMAL	Quest Electronics	1
DRAWER SLIDE: BALL BEARING, 13" L	ETCO	1
DRIVE: ANGLE, MINIATURE, DUAL SHAFT	Small Parts	2
DUMMY LOAD:	See Load:	
FADER: STEREO	See Resistor: Variable, Fade	
FAN BLADE: SMALL	ETCO	2
FAN: 12 V DC	Semiconductors Surplus	1
FAN: EQUIPMENT BLOWER	ETCO	3
FAN: EQUIPMENT BLOWER	Fair Radio Sales	10
FAN: MUFFIN, 115 VOLT, ROTRON	Fair Radio Sales	2
FAN: MUFFIN, 115 VOLT, ROTRON	MHZ Electronics	1
FAN: MUFFIN, 115 VOLT, ROTRON	Surplus Electronics	1
FAN: MUFFIN, 220 V AC, IMC BOXER	Fair Radio Sales	1
FAN: MUFFIN, BISCUIT, BOXER, TUNNEL	ETCO	5
FAN: MUFFIN, BISCUIT,FEATHER,ROTRON	Semiconductors Surplus	3
FAN: MUFFIN, BOXER, 115 VAC	Jameco Electronics	2
FAN: MUFFIN, MINIATURE, ROTARY	Mouser Electronics	1
FAN: OPEN FRAME, 115 & 230 VAC	Surplus Electronics	2
FAN: RACK PANEL MOUNTED, DUAL, 120V	Atlantic Surplus Sales	1
FAN: SQUIRREL CAGE	Semiconductors Surplus	1
FAN: SQUIRREL CAGE, 28 V DC	Fair Radio Sales	1
FEET:	See Hardware	
FELT: SHEET, 1/8 TO 3/8" THICK	Small Parts	6

COMPONENT	COMPANY NAME	VARIETY STOCKED
FERRITE: 'E' CORES	Amidon Associates	6
FERRITE: BEAD	Semiconductors Surplus	5
FERRITE: BEAD	Circuit Specialists	1
FERRITE: BEAD	Amidon Associates	6
FERRITE: BEAD	Palomar Engineers	3
FERRITE: BEAD	Radiokit	9
FERRITE: BEAD	Digital Research	1
FERRITE: BEAD	BCD Radio Parts	1
FERRITE: BEAD, 2-HOLE	Radiokit	1
FERRITE: BEAD, 6-HOLE	Radiokit	1
FERRITE: BEAD, 6-HOLE	Circuit Specialists	1
FERRITE: BEAD, 6-HOLE	Palomar Engineers	1
FERRITE: BEAD, JW MILLER	Mouser Electronics	10
FERRITE: CORE, TOROIDAL	See Core: Toroidal	
FERRITE: POT CORE	Amidon Associates	6
FERRITE: ROD	Palomar Engineers	1
FERRITE: ROD	Amidon Associates	4
FERRITE: ROD	Radiokit	4
FERRITE: ROD	Semiconductors Surplus	1
FERRITE: ROD	Digital Research	1
FERRITE: ROD	BCD Radio Parts	1
FERRITE: ROD	ETCO	7
FERRITE: ROD, JW MILLER	Mouser Electronics	10
FERRITE: SLEEVE	Amidon Associates	1
FIBER OPTIC LIGHT GUIDE:	Edmund Scientific	29
FILAMENT: LACING, WAXED	Mouser Electronics	3
FILTER CHOKE:	See Inductor: Filter Choke	
FILTER: ALTERNATOR NOISE	ETCO	1
FILTER: ATV 426, 438 MHZ, BP	Spectrum International	2
FILTER: BAND PASS, COAXIAL, MICROWV	Lectronic	
FILTER: CERAMIC, 10.7 MHZ	Circuit Specialists	3
FILTER: CERAMIC, 455 KHZ, PC MOUNT	Surplus Electronics	1
FILTER: CERAMIC, CLEVITE, 455 KHZ	MHZ Electronics	2
FILTER: CERAMIC, MALLORY, 10.7 MHZ	Mouser Electronics	1
FILTER: CERAMIC, MALLORY, 455 KHZ	Mouser Electronics	2
FILTER: CERAMIC, MATSUSHIRA, 455KHZ	MHZ Electronics	1
FILTER: CERAMIC, MURATA, 10.7 MHZ	MHZ Electronics	2
FILTER: CERAMIC, MURATA, 10.7 MHZ	Semiconductors Surplus	1
FILTER: CERAMIC, MURATA, 455 KHZ	Semiconductors Surplus	4
FILTER: CERAMIC, MURATA, 455 KHZ	MHZ Electronics	9
FILTER: CERAMIC, NIPPON, 455 KHZ	MHZ Electronics	3
FILTER: CERAMIC, TOKIN, 455 KHZ	MHZ Electronics	1
FILTER: COLLINS, 250 HZ	Fuji-Svea	1
FILTER: COLLINS, 250 HZ	Fox Tango Corporation	1
FILTER: COLLINS, NOTCH, Q-MULT	Barker & Williamson	1
FILTER: CRYSTAL, 10.7 MHZ	Piezo Technology	24
FILTER: CRYSTAL, 10.7 MHZ	Spectrum International	6
FILTER: CRYSTAL, 21.4 MHZ	Piezo Technology	21
FILTER: CRYSTAL, 41 MHZ	Spectrum International	1
FILTER: CRYSTAL, 45 MHZ, 15 KHZ BW	Piezo Technology	1
FILTER: CRYSTAL, 9 MHZ	Spectrum International	12
FILTER: CRYSTAL, ATLAS	Semiconductors Surplus	6
FILTER: CRYSTAL, ATLAS	MHZ Electronics	7
FILTER: CRYSTAL, CD, 45 MHZ	MHZ Electronics	1

COMPONENT	COMPANY NAME	VARIETY STOCKED
FILTER: CRYSTAL, CUSTOM	Piezo Technology	
FILTER: CRYSTAL, DRAKE	Fox Tango Corporation	9
FILTER: CRYSTAL, DRAKE	Sherwood Engineering	10
FILTER: CRYSTAL, HEATH	Fuji-Svea	4
FILTER: CRYSTAL, HEATH	Fox Tango Corporation	4
FILTER: CRYSTAL, KENWOOD	Fuji-Svea	8
FILTER: CRYSTAL, KENWOOD	Fox Tango Corporation	9
FILTER: CRYSTAL, MOTOROLA, 11.7 MHZ	MHZ Electronics	1
FILTER: CRYSTAL, NIKKO, 7.8 MHZ	MHZ Electronics	1
FILTER: CRYSTAL, PTI, 12 & 21.4 MHZ	MHZ Electronics	2
FILTER: CRYSTAL, TYCO, 10.7 MHZ	Semiconductors Surplus	1
FILTER: CRYSTAL, TYCO/CD, 10.7 MHZ	MHZ Electronics	1
FILTER: CRYSTAL, YAESU	Fuji-Svea	23
FILTER: CRYSTAL, YAESU	Fox Tango Corporation	26
FILTER: DTMF	See IC	
FILTER: EFCL455K13E, CRYSTAL	Semiconductors Surplus	1
FILTER: EFCL455K40B2, CRYSTAL	Semiconductors Surplus	1
FILTER: FHA103-4, CRYSTAL, 10.7 MHZ	Semiconductors Surplus	1
FILTER: FX-07800L, CRYSTAL, 7.8 MHZ	Semiconductors Surplus	1
FILTER: HIGH PASS, 50 MHZ CUT-OFF	ETCO	1
FILTER: IGNITION NOISE	ETCO	1
FILTER: LC, 500 KHZ, COLLINS	Fair Radio Sales	1
FILTER: LOW PASS, 200 MHZ CUT OFF	Spectrum International	2
FILTER: LOW PASS, COAXIAL, MICROWV	Lectronic	
FILTER: MECHANICAL, COLLINS, 455KHZ	MHZ Electronics	1
FILTER: MECHANICAL, COLLINS, USED	Fair Radio Sales	9
FILTER: MECHANICAL, COLLINS, USED	Star-Tronics	1
FILTER: MECHANICAL, F-307 MOTOROLA	Fair Radio Sales	2
FILTER: MECHANICAL, KOKUSAI, 455KHZ	MHZ Electronics	1
FILTER: MICROWAVE, COAX, LO-,HI-,BP	Lectronic	
FILTER: OPTICAL DISPLAY	See Bezel Filter:	
FILTER: PRESELECT BP, 1296,1691 MHZ	Spectrum International	2
FILTER: PRESELECT BP, 144, 432 MHZ	Spectrum International	2
FILTER: RF, COAXIAL, LW-P, HI-P, BP	Fuji-Svea	
FILTER: RF, VHF LOW PASS, 170 MHZ	Aldelco	1
FILTER: TEW, 10.6935 MHZ	MHZ Electronics	1
FILTER: WAVEGUIDE, LW-P, HI-P, BP	Lectronic	
FLOAT: STAINLESS STEEL	Small Parts	5
FLOW INDICATOR: GASOLINE FUEL FLOW	Star-Tronics	1
FOAM: ANTISTATIC	Quest Electronics	2
FOAM: ANTISTATIC	Jameco Electronics	1
FOIL: CONDUCTIVE, ADHESIVE, 3/8" W	ETCO	1
FREQUENCY DOUBLER: RF, TO 1 GHZ IN	Mini-Circuits Laboratory	11
FREQUENCY METER:	See Also Wavemeter:	
FREQUENCY METER: COAXIAL, MICROWAVE	Lectronic	
FREQUENCY MULTIPLIER: WAVEGUIDE	Lectronic	
FUSE:	BCD Radio Parts	7
FUSE: 1/100 & 1/200 AMP	Optoelectronics	2
FUSE: 3AG, 1 AMP	Semiconductors Surplus	1
FUSE: 3AG, FAST ACTING	Fuji-Svea	16
FUSE: 3AG, NORMAL-BLO	Digi-Key	29

COMPONENT	COMPANY NAME	VARIETY STOCKED
FUSE: 3AG, NORMAL-BLO, PIGTAIL	Digi-Key	7
FUSE: 3AG, PICO, AXIAL LEAD	Active Electronics	18
FUSE: 3AG, SLO-BLO	Digi-Key	43
FUSE: 3AG, SLO-BLO, 1/100-30 AMP	Active Electronics	41
FUSE: 3AG, SLOW-BLO, PIGTAIL	Digi-Key	16
FUSE: 3AG, STANDARD, 1/16-5 AMP	Active Electronics	21
FUSE: ABC, CERAMIC	Fuji-Svea	6
FUSE: ABC, CERAMIC, 1/4x1.25"	Mouser Electronics	10
FUSE: AGC, FAST ACTING	Jameco Electronics	6
FUSE: AGC, FAST ACTING	Fuji-Svea	6
FUSE: AGX, FAST ACT, INSTRUMENT	Fuji-Svea	6
FUSE: CHEMICAL, BEL FUSE, TV	Fuji-Svea	7
FUSE: FAST ACTING	Daytapro Electronics	9
FUSE: FAST ACTING, INSTRUMENT	Mouser Electronics	8
FUSE: FAST ACTING, STANDARD	Mouser Electronics	20
FUSE: MDC 4	Key Electronics	1
FUSE: MDL, SLOW BLOWING	Fuji-Svea	19
FUSE: MDL, SLOW BLOWING	Jameco Electronics	8
FUSE: MIDGET, 20x5mm	Star-Tronics	5
FUSE: MIDGET, 20x5mm, FAST ACTING	Mouser Electronics	5
FUSE: PIG TAIL	Fuji-Svea	8
FUSE: PIG TAIL, 1 A	Daytapro Electronics	1
FUSE: PIG TAIL, 1/2 A	Mouser Electronics	1
FUSE: RF, COAXIAL, 50 OHM, -1.5 GHZ	Elcom Systems	2
FUSE: RF, COAXIAL, 75 OHM, -1.5 GHZ	Elcom Systems	1
FUSE: SLOW BLOWING	Daytapro Electronics	7
FUSE: SLOW BLOWING	Mouser Electronics	20
FUSE: SUBMINIATURE, .769"	Fuji-Svea	4
FUSEHOLDER: BLOCK, .25" DIA FUSES	Digi-Key	4
FUSEHOLDER: BLOCK, 1 & 1 1/4" FUSE	Fuji-Svea	2
FUSEHOLDER: BLOCK, MIDGET BASE FUSE	Fuji-Svea	1
FUSEHOLDER: BLOCK, MOLDED BAKELITE	Mouser Electronics	5
FUSEHOLDER: BLOCK, PHENOLIC	Daytapro Electronics	1
FUSEHOLDER: CLIP, PC BOARD	Digi-Key	1
FUSEHOLDER: CLIP, SOLDER & PC	Mouser Electronics	6
FUSEHOLDER: CLIP, SOLDER & PC	Active Electronics	4
FUSEHOLDER: IN-LINE, HEAVY DUTY	Digi-Key	1
FUSEHOLDER: IN-LINE, STANDARD	BCD Radio Parts	1
FUSEHOLDER: IN-LINE, STANDARD	Digital Research	2
FUSEHOLDER: IN-LINE, STANDARD	Fuji-Svea	1
FUSEHOLDER: IN-LINE, STANDARD	Mouser Electronics	7
FUSEHOLDER: IN-LINE, TRIPLE, PLUG	BCD Radio Parts	1
FUSEHOLDER: PANEL MOUNT	Daytapro Electronics	2
FUSEHOLDER: PANEL MOUNT	Jameco Electronics	1
FUSEHOLDER: PANEL MOUNT	Semiconductors Surplus	1
FUSEHOLDER: PANEL MOUNT	Fuji-Svea	4
FUSEHOLDER: PANEL MOUNT	Active Electronics	2
FUSEHOLDER: PANEL MOUNT	Digi-Key	1
FUSEHOLDER: PANEL MOUNT	Mouser Electronics	15
FUSEHOLDER: PANEL MOUNT	BCD Radio Parts	1
FUSEHOLDER: PANEL MT, MIDGET FUSE	Fuji-Svea	1
FUSEHOLDER: TWIN-CLIP, BUSS	Mouser Electronics	2
FUSIBLE METAL: LOW MELTING POINT	Small Parts	4
GAS SENSOR ELEMENT: COMBUSTABLE GAS	ETCO	1
GEAR: RACK & PINIONS, BRASS	Small Parts	13

COMPONENT	COMPANY NAME	VARIETY STOCKED
GEAR: SPUR, MOLDED DELRIN	Small Parts	38
GEAR: SPUR, NYLON, BRASS, STEEL	Edmund Scientific	ASST
GEAR: TELETYPE	Atlantic Surplus Sales	33
GEAR: TELETYPE	Typetronics	
GENERATOR: FMA5 AUDIO SUBCARRIER	P.C. Electronics	1
GENERATOR: HAND CRANK, 28 & 300 V	Fair Radio Sales	1
GENERATOR: TELEPHONE, CRANK, 100 V	Edmund Scientific	1
GRIPPER STRIP:	See Also: Velcro:	
GRIPPER STRIP: HOOK & LOOP FASTENER	Small Parts	8
GROMMET:	See Hardware: Grommet	
GROUND STRAP: COPPER, 3/16 & 3/8" W	Nemal Electronics	3
GYRO: RATE, NORTHRUP CN-1278A	Star-Tronics	1
HANDLE: BALL, CYLINDRICAL, 3-ARM	Small Parts	18
HANDLE: PANEL	Daytapro Electronics	5
HANDLE: RADIO & TV, SONY	BCD Radio Parts	1
HARDWARE: ANCHOR, GUYING, EVE, WALL	Ham Hardware Headquarters	
HARDWARE: BOLT, CLOSED EYE, SS	Ham Hardware Headquarters	
HARDWARE: BOLT, HEX HEAD, NYLON	Small Parts	32
HARDWARE: BOLT, HEX HEAD, SS	Small Parts	58
HARDWARE: BOLT, HEX HEAD, SS	Ham Hardware Headquarters	
HARDWARE: BOLT, HEX HEAD, TITANIUM	Small Parts	7
HARDWARE: BRACKET, MOUNTING, SMALL	Daytapro Electronics	4
HARDWARE: BRACKET, PRINTED CKT BRD	Quest Electronics	1
HARDWARE: BRACKET, VOLUME CONTRL MT	ETCO	1
HARDWARE: BUSHING, BAKELITE	Daytapro Electronics	3
HARDWARE: COTTER PIN, SS	Small Parts	19
HARDWARE: EJECTOR, CIRCUIT BOARD	Sintec	4
HARDWARE: FEET	ETCO	5
HARDWARE: FEET, POLY, ADHESIVE	Mouser Electronics	8
HARDWARE: FEET, RUBBER	Surplus Electronics	2
HARDWARE: FEET, RUBBER, HARD	BCD Radio Parts	1
HARDWARE: FEET, RUBBER, SCREW MT	Mouser Electronics	5
HARDWARE: FEET, RUBBER, SNAP-IN	Daytapro Electronics	5
HARDWARE: FEET, RUBBER, STICK ON	Daytapro Electronics	4
HARDWARE: FEET, STICK ON, 3-M	Semiconductors Surplus	2
HARDWARE: FEET, VINYL, SCREW MT	Daytapro Electronics	3
HARDWARE: FEET, VINYL, SCREW MT	Mouser Electronics	4
HARDWARE: GROMMET, PLASTIC	Mouser Electronics	8
HARDWARE: GROMMET, RUBBER	Daytapro Electronics	5
HARDWARE: GROMMET, RUBBER	Mouser Electronics	6
HARDWARE: GUIDE, PRINTED CKT BOARD	Quest Electronics	2
HARDWARE: GUIDE, PRINTED CKT BOARD	Sintec	4
HARDWARE: GUIDE, PRINTED CKT BOARD	Daytapro Electronics	1
HARDWARE: GUIDE, PRINTED CKT BOARD	Digi-Key	2
HARDWARE: GUIDE, RACK CARD	Daytapro Electronics	2
HARDWARE: HITCH PIN, SPRING STEEL	Small Parts	7
HARDWARE: HOLE PLUG, SOLID, METAL	Mouser Electronics	8
HARDWARE: HOLE PLUG, SOLID, NYLON	Mouser Electronics	7
HARDWARE: HOLE PLUG, VENTED, NYLON	Mouser Electronics	2
HARDWARE: INSERT, THREADED, E-Z LOK	Small Parts	7
HARDWARE: LUG, SOLDER	Daytapro Electronics	3
HARDWARE: LUG, SOLDER	Radiokit	2

COMPONENT	COMPANY NAME	VARIETY STOCKED
HARDWARE: LUG, SOLDER, LOCKING	Mouser Electronics	25
HARDWARE: NUTS, ACORN, BRASS	Ham Hardware Headquarters	
HARDWARE: NUTS, CAP, BRASS, NP	Small Parts	5
HARDWARE: NUTS, CAP, NYLON	Small Parts	5
HARDWARE: NUTS, CAP, SS	Small Parts	6
HARDWARE: NUTS, CAP, SS	Ham Hardware Headquarters	
HARDWARE: NUTS, CAPTIVE, PANEL EDGE	Star-Tronics	1
HARDWARE: NUTS, HEX, NYLON, METRIC	Small Parts	6
HARDWARE: NUTS, HEXAGON	Mouser Electronics	5
HARDWARE: NUTS, HEXAGON	Daytapro Electronics	3
HARDWARE: NUTS, HEXAGON	Radiokit	2
HARDWARE: NUTS, HEXAGON, METRIC	Fuji-Svea	ASST
HARDWARE: NUTS, HEXAGON, METRIC	Small Parts	6
HARDWARE: NUTS, HEXAGON, NYLON	Small Parts	9
HARDWARE: NUTS, HEXAGON, NYLON	Daytapro Electronics	3
HARDWARE: NUTS, HEXAGON, SS	Small Parts	13
HARDWARE: NUTS, HEXAGON, SS, LARGE	Ham Hardware Headquarters	
HARDWARE: NUTS, HEXAGON, TITANIUM	Small Parts	1
HARDWARE: NUTS, JAM, TITANIUM, 1/4"	Small Parts	1
HARDWARE: NUTS, LOCK, METRIC	Small Parts	4
HARDWARE: NUTS, LOCK, NYLON	Small Parts	6
HARDWARE: NUTS, LOCK, SS, NYLON INS	Small Parts	9
HARDWARE: NUTS, POTENTIOMETER	Mouser Electronics	5
HARDWARE: NUTS, POTENTIOMETER	Alpha Electronic Labs	
HARDWARE: NUTS, POTENTIOMETER	ETCO	3
HARDWARE: NUTS, ROTARY SWITCH	Mouser Electronics	3
HARDWARE: NUTS, SELF-CLENCHING, HEX	Ham Hardware Headquarters	
HARDWARE: NUTS, SELF-CLINCHING	Small Parts	12
HARDWARE: NUTS, SQUARE, SS	Ham Hardware Headquarters	
HARDWARE: NUTS, SWITCH	ETCO	3
HARDWARE: NUTS, SWITCH	Mouser Electronics	10
HARDWARE: NUTS, THUMB, BRASS, BATRY	Ham Hardware Headquarters	
HARDWARE: NUTS, THUMB, BRASS, NP	Small Parts	5
HARDWARE: NUTS, THUMB, NYLON	Small Parts	5
HARDWARE: NUTS, WING, BRASS	Ham Hardware Headquarters	
HARDWARE: NUTS, WING, NYLON	Small Parts	8
HARDWARE: NUTS, WING, SS	Small Parts	4
HARDWARE: NUTS, WING, SS	Ham Hardware Headquarters	
HARDWARE: RETAINING RING, INT & EXT	Small Parts	34
HARDWARE: RIVET, DRIVE, ALUMINUM	Small Parts	9
HARDWARE: SCREW, BIND HEAD, PHILLIP	Star-Tronics	1
HARDWARE: SCREW, BINDING HEAD	Daytapro Electronics	9
HARDWARE: SCREW, BINDING HEAD, SS	Small Parts	82
HARDWARE: SCREW, BINDING HEAD,NYLON	Small Parts	45
HARDWARE: SCREW, CAP SOCKET, METRIC	Small Parts	19
HARDWARE: SCREW, CAPTIVE, PANEL	ETCO	1
HARDWARE: SCREW, CHEESE HEAD,METRIC	Small Parts	37
HARDWARE: SCREW, FILLISTER HEAD, SS	Small Parts	82
HARDWARE: SCREW, FLAT HEAD	Mouser Electronics	6
HARDWARE: SCREW, FLAT HEAD, NYLON	Small Parts	32
HARDWARE: SCREW, FLAT HEAD, SS	Small Parts	82
HARDWARE: SCREW, FLAT PHILLIPS HEAD	Small Parts	37
HARDWARE: SCREW, HEX HEAD, NYLON	Small Parts	45
HARDWARE: SCREW, HEX HEAD, SS	Small Parts	16
HARDWARE: SCREW, HIGH RATE HELIX	Small Parts	1
HARDWARE: SCREW, MACHINE, METRIC	Fuji-Svea	ASST
HARDWARE: SCREW, MACHINE, SS	Ham Hardware Headquarters	
HARDWARE: SCREW, NYLON	Daytapro Electronics	7
HARDWARE: SCREW, NYLON, METRIC	Small Parts	49

COMPONENT	COMPANY NAME	VARIETY STOCKED
HARDWARE: SCREW, OVAL HEAD, METRIC	Small Parts	13
HARDWARE: SCREW, PAN HEAD, NYLON	Small Parts	8
HARDWARE: SCREW, PAN PHILLIPS HEAD	Small Parts	37
HARDWARE: SCREW, ROUND HEAD	Radiokit	10
HARDWARE: SCREW, ROUND HEAD	Mouser Electronics	21
HARDWARE: SCREW, ROUND HEAD, SLOT	Star-Tronics	6
HARDWARE: SCREW, SELF-TAPPING, SS	Small Parts	19
HARDWARE: SCREW, SHEET METAL	Mouser Electronics	13
HARDWARE: SCREW, SHEET METAL TAP	Ham Hardware Headquarters	
HARDWARE: SCREW, SOCKET CAP HEAD,SS	Small Parts	33
HARDWARE: SCREW, THUMB HEAD, BRASS	Aldelco	3
HARDWARE: SCREW, THUMB HEAD, BRASS	Small Parts	19
HARDWARE: SCREW, THUMB HEAD, NYLON	Small Parts	15
HARDWARE: SCREW-EYE, ALUMINUM, 6/32	Skylane Products	1
HARDWARE: SET SCREW, COMB HD, NYLON	Small Parts	20
HARDWARE: SET SCREW, SOCKET HEAD,SS	Small Parts	24
HARDWARE: SET SCREW, SOCKET, METRIC	Small Parts	13
HARDWARE: SHOCK CUSHIONING, VLIER	Small Parts	1
HARDWARE: SHOCK MOUNT, AUDIO ISOL	ETCO	1
HARDWARE: SHOCK MOUNT, RUBBER	Fair Radio Sales	7
HARDWARE: SPACER, ALUMINUM	Radiokit	2
HARDWARE: SPACER, ALUMINUM, ASSORT	Circuit Specialists	1
HARDWARE: SPACER, CERAMIC, THREADED	Radiokit	1
HARDWARE: SPACER, HEX, ALUM, THRED	Small Parts	17
HARDWARE: SPACER, HEX, ALUM, THRED	Digi-Key	8
HARDWARE: SPACER, HEX, ALUMINUM	Daytapro Electronics	4
HARDWARE: SPACER, HEX, BRASS, THRED	Mouser Electronics	8
HARDWARE: SPACER, HEX, PLASTIC	Daytapro Electronics	3
HARDWARE: SPACER, HEX, PVDC DIELECT	Mouser Electronics	15
HARDWARE: SPACER, INSULATOR	See Also Insulator: Spacer	
HARDWARE: SPACER, METAL, THREADED	Digi-Key	4
HARDWARE: SPACER, METAL, THREADED	Digital Research	1
HARDWARE: SPACER, NYLON, ASSORTMENT	Circuit Specialists	1
HARDWARE: SPACER, NYLON, THREADED	Digital Research	1
HARDWARE: SPACER, ROUND, ALUMINUM	Daytapro Electronics	4
HARDWARE: SPACER, ROUND, ALUMINUM	Mouser Electronics	18
HARDWARE: SPACER, ROUND, ALUMINUM	Digi-Key	4
HARDWARE: SPACER, ROUND, BRASS,THRD	Mouser Electronics	8
HARDWARE: SPACER, ROUND, NYLON	Small Parts	36
HARDWARE: SPACER, ROUND, PLASTIC	Daytapro Electronics	6
HARDWARE: SPACER, ROUND, PVDC DIEL	Mouser Electronics	12
HARDWARE: SPLICE, WIRE, BUTT	Active Electronics	3
HARDWARE: SPLICE, WIRE, BUTT	Mouser Electronics	2
HARDWARE: SPLICE, WIRE, SCOTCH LOK	BCD Radio Parts	1
HARDWARE: SPRING PIN, HOLLOW, STEEL	Small Parts	29
HARDWARE: STAND OFF, RIVET MT, THRD	Star-Tronics	1
HARDWARE: STAPLE, WIRE, FIBER INSUL	Aldelco	1
HARDWARE: STEEL CABLE ENDS	Small Parts	15
HARDWARE: STRAIN RELIEF, POWER CORD	Mouser Electronics	10
HARDWARE: STRAIN RELIEF, POWER CORD	Digital Research	1
HARDWARE: STRAIN RELIEF, POWER CORD	BCD Radio Parts	1
HARDWARE: STRAIN RELIEF, POWER CORD	Daytapro Electronics	3
HARDWARE: STRAIN RELIEF, POWER CORD	ETCO	1
HARDWARE: STRAP, CABLE, BLIND HOLE	Mouser Electronics	3
HARDWARE: TERMINAL, FASHION RECPT	Active Electronics	2
HARDWARE: TERMINAL, HOOK TOUNGE	Mouser Electronics	4
HARDWARE: TERMINAL, LOCKWASHER	ETCO	8
HARDWARE: TERMINAL, RING	Star-Tronics	1
HARDWARE: TERMINAL, RING	Key Electronics	1

COMPONENT	COMPANY NAME	VARIETY STOCKED
HARDWARE: TERMINAL, RING	ETCO	7
HARDWARE: TERMINAL, RING	Active Electronics	22
HARDWARE: TERMINAL, RING	Mouser Electronics	6
HARDWARE: TERMINAL, SPADE TONGUE	Fair Radio Sales	1
HARDWARE: TERMINAL, SPADE TONGUE	ETCO	7
HARDWARE: TERMINAL, SPADE TONGUE	Active Electronics	16
HARDWARE: TERMINAL, SPADE TOUNGE	Mouser Electronics	18
HARDWARE: TERMINAL, SPADE TOUNGE	Daytapro Electronics	3
HARDWARE: TERMINAL, WINDOW SPLICE	Active Electronics	3
HARDWARE: TURNBUCKLE, HVY DUTY,GALV	Ham Hardware Headquarters	
HARDWARE: TURNBUCKLE, SS	Ham Hardware Headquarters	
HARDWARE: U-BOLT, GUYING CABLE TYPE	Ham Hardware Headquarters	
HARDWARE: U-BOLT, SS	Ham Hardware Headquarters	
HARDWARE: WASHER, COUNTERSUNK	Mouser Electronics	3
HARDWARE: WASHER, FIBER, FLAT	Radiokit	1
HARDWARE: WASHER, FIBER, FLAT	ETCO	1
HARDWARE: WASHER, FLAT	Daytapro Electronics	3
HARDWARE: WASHER, FLAT	Mouser Electronics	5
HARDWARE: WASHER, FLAT	Radiokit	2
HARDWARE: WASHER, FLAT, FIBRE	Mouser Electronics	6
HARDWARE: WASHER, FLAT, METRIC	Small Parts	6
HARDWARE: WASHER, FLAT, NYLON	Small Parts	21
HARDWARE: WASHER, FLAT, NYLON,METRC	Small Parts	6
HARDWARE: WASHER, FLAT, SS	Small Parts	21
HARDWARE: WASHER, FLAT, SS, LARGE	Ham Hardware Headquarters	
HARDWARE: WASHER, FLAT, TITANIUM	Small Parts	2
HARDWARE: WASHER, INTERNAL TOOTH	Mouser Electronics	4
HARDWARE: WASHER, LOCK	Mouser Electronics	4
HARDWARE: WASHER, LOCK TOOTH, INT	Daytapro Electronics	3
HARDWARE: WASHER, LOCK, #3, SS	Star-Tronics	1
HARDWARE: WASHER, LOCK, SS	Small Parts	11
HARDWARE: WASHER, LOCK, STAR	Radiokit	2
HARDWARE: WASHER, LOCK, TITANIUM	Small Parts	1
HARDWARE: WASHER, LOCK, TOOTH, SS	Ham Hardware Headquarters	
HARDWARE: WASHER, POTENTIOMETER	Alpha Electronic Labs	
HARDWARE: WASHER, ROTARY SWITCH	Mouser Electronics	3
HARDWARE: WASHER, SHOULDER, FIBRE	Mouser Electronics	5
HARDWARE: WASHER, SPLIT SPRING, SS	Ham Hardware Headquarters	
HARDWARE: WIRE COUPLER, INSULATED	Mouser Electronics	4
HEAD: 2-TRACK, RECORD	ETCO	1
HEAD: 8-TRACK, PLAY, STEREO	ETCO	1
HEAD: 8-TRACK, RECORD, PLAY	Fuji-Svea	1
HEAD: CASSETTE, RECORD, PLAY	Semiconductors Surplus	1
HEAD: CASSETTE, RECORD, PLAY, 2 TRK	Fuji-Svea	2
HEAD: CASSETTE, RECORD, PLAY, MONO	Fuji-Svea	1
HEADER COVER:	Sintec	28
HEADER COVER:	Jameco Electronics	3
HEADER JUMPER: DIPATCH	Digi-Key	2
HEADER STRIP SOCKET: SIP	Digi-Key	4
HEADER STRIP:	Mouser Electronics	2
HEADER STRIP: BERG STICK	Digital Research	1
HEADER STRIP: DOUBLE ROW	BCD Radio Parts	1
HEADER STRIP: DOUBLE ROW	Jameco Electronics	6
HEADER STRIP: DOUBLE ROW	Mouser Electronics	2

COMPONENT	COMPANY NAME	VARIETY STOCKED
HEADER STRIP: DOUBLE ROW	Quest Electronics	10
HEADER STRIP: DOUBLE ROW	Circuit Specialists	10
HEADER STRIP: FEMALE	Digi-Key	2
HEADER STRIP: MALE	Digi-Key	38
HEADER STRIP: PROTECTED, LATCHES	Digi-Key	48
HEADER STRIP: SIP, FORKED CONTACT	Digi-Key	1
HEADER STRIP: SOLDER	Quest Electronics	4
HEADER STRIP: WIRE WRAP	Quest Electronics	4
HEADER:	See Also Dip Plug:	
HEADER: DIP	Alpha Electronic Labs	5
HEADER: DIP	Digital Research	1
HEADER: DIP	Digi-Key	4
HEADER: DIP	Babylon Electronics	1
HEADER: DIP	Circuit Specialists	4
HEADER: DIP	Quest Electronics	6
HEADER: DIP	Fuji-Svea	2
HEADER: DIP	Mouser Electronics	23
HEADER: DIP	Jameco Electronics	3
HEADER: DIP, FORK & POST CONTACT	Sintec	28
HEADER: DIP, GOLD PIN	Electronic Marketplace	4
HEADER: DIP, PROGRAMMABLE SHUNTS	Digi-Key	3
HEADER: DIP, PROGRAMMABLE SHUNTS	Sintec	10
HEADER: DIP, SINGLE ROW	Sintec	2
HEADSET CUSHION: CHAMOIS	Fair Radio Sales	1
HEADSHELL: RECORDPLAYER CARTRIDGE	Fuji-Svea	1
HEADSHELL: RECORDPLAYER CARTRIDGE	ETCO	2
HEAT ELEMENT: COIL, 1000 W, 120 VAC	ETCO	1
HEAT ELEMENT: COIL, 1000 W, 220 VAC	ETCO	1
HEAT ELEMENT: PLATE, TEMP CONTROLLD	BCD Radio Parts	1
HEAT ELEMENT: PLATE, TEMP CONTROLLD	Digital Research	1
HEAT ELEMENT: PLATE, TEMP CONTROLLD	ETCO	1
HEAT ELEMENT: ROPE, 100 W, 120 VAC	ETCO	1
HEATSINK: DIODE RECTIFIER MT	Fair Radio Sales	4
HEATSINK: DIODE RECTIFIER MT	Jameco Electronics	1
HEATSINK: DUAL TO-36 MT	Fair Radio Sales	1
HEATSINK: HS- SERIES, AAVID	Digi-Key	32
HEATSINK: POWER SEMICONDUCTORS	Daytapro Electronics	4
HEATSINK: TO-202	Jameco Electronics	1
HEATSINK: TO-220	Key Electronics	1
HEATSINK: TO-220	Jameco Electronics	2
HEATSINK: TO-220	BCD Radio Parts	2
HEATSINK: TO-220	Semiconductors Surplus	1
HEATSINK: TO-220	Active Electronics	7
HEATSINK: TO-220	Digital Research	3
HEATSINK: TO-66	Surplus Electronics	1
HEATSINK: TO-66	Active Electronics	2
HEATSINK: TO-92	Key Electronics	1
HEATSINK: TRANSISTOR	Quest Electronics	4
HEATSINK: TRANSISTOR, TO-3	Active Electronics	7
HEATSINK: TRANSISTOR, TO-3	ETCO	4
HEATSINK: TRANSISTOR, TO-3	Key Electronics	2
HEATSINK: TRANSISTOR, TO-3	Jameco Electronics	4
HEATSINK: TRANSISTOR, TO-3	Surplus Electronics	3
HEATSINK: TRANSISTOR, TO-3	Fair Radio Sales	5

COMPONENT	COMPANY NAME	VARIETY STOCKED
HEATSINK: TRANSISTOR, TO-3, LARGE	BCD Radio Parts	1
HEATSINK: TRANSISTOR, TO-3, LARGE	Digital Research	1
HEATSINK: TRANSISTOR, TO-5	Jameco Electronics	1
HEATSINK: TRANSISTOR, TO-5	BCD Radio Parts	1
HEATSINK: TRANSISTOR, TO-5, CLIP-ON	Active Electronics	1
HEATSINK: TRANSISTOR, TO-5, DUAL	Digital Research	1
HEATSINK: VERY LARGE ASSEMBLY	Fair Radio Sales	2
HINGE: PIANO TYPE, 8 1/2"	ETCO	1
HINGE: PIANO TYPE, STAINLESS STEEL	Small Parts	5
HYBRID TEE: COAXIAL, MICROWAVE	Lectronic	
IC:	ETCO	84
IC:	BCD Radio Parts	28
IC:	Digital Research	39
IC: 1173AN-1, MUSICAL MPU	Jameco Electronics	1
IC: 11C24 TTL VCM	Active Electronics	1
IC: 11C44 PHASE FREQUENCY DETECTOR	Active Electronics	1
IC: 11C90, 650 MHZ PRESCALER, ECL	Active Electronics	1
IC: 11C90, 650 MHZ PRESCALER, ECL	Circuit Specialists	1
IC: 11C90, 650 MHZ PRESCALER, ECL	Aldelco	1
IC: 11C90, 650 MHZ PRESCALER, ECL	Jameco Electronics	1
IC: 11C90, 650 MHZ PRESCALER, ECL	Quest Electronics	1
IC: 1404 SHIFT REGISTER	Active Electronics	1
IC: 1408L-8 D/A CONVERTER	Active Electronics	1
IC: 1412 VOLTAGE REGULATOR	B.G. Micro	1
IC: 1458 DUAL 741, 8 PIN DIP	Digital Research	1
IC: 1469 VOLTAGE REGULATOR	Digital Research	1
IC: 1488, 1489, INTERFACE	Digi-Key	3
IC: 1488, 1489, INTERFACE	Jameco Electronics	2
IC: 1488, 1489, INTERFACE	Key Electronics	2
IC: 1488, 1489, INTERFACE	Electronic Marketplace	2
IC: 1488, 1489, INTERFACE, RS-232	Babylon Electronics	2
IC: 1595 MULTIPLIER, 4 QUADRANT	Babylon Electronics	1
IC: 1602 UART	BCD Radio Parts	1
IC: 1602 UART	B.G. Micro	1
IC: 1771 FLOPPY DISK CONTROLLER, SD	B.G. Micro	1
IC: 1791 FLOPPY DISK CONTROLLER, DS	B.G. Micro	1
IC: 17XX FLOPPY DISK CONTROLLER	BCD Radio Parts	2
IC: 17XX FLOPPY DISK CONTROLLER	Quest Electronics	4
IC: 17XX FLOPPY DISK CONTROLLER	Alpha Electronic Labs	4
IC: 17XX FLOPPY DISK CONTROLLER	Active Electronics	5
IC: 17XX SERIES	Electronic Marketplace	2
IC: 1800 SERIES, SPECIAL FUNCTION	Digi-Key	6
IC: 1802 SERIES MICROPROCESSOR	Quest Electronics	11
IC: 1802 SERIES MICROPROCESSOR	Active Electronics	23
IC: 1802 SERIES MICROPROCESSOR	Jameco Electronics	1
IC: 1M640X UART	Quest Electronics	2
IC: 2001 SUBSCRIBER LINE INTERFACE	ITT Microsystems	1
IC: 2002 SUBSCRIBER LINE INTERFACE	ITT Microsystems	1
IC: 2111 FM DETECTOR, SPRAGUE	Digital Research	1
IC: 2416 CHARGE COUPLED DEVICE, CCD	Babylon Electronics	1
IC: 2500 SERIES, SHIFT REGISTER	Quest Electronics	9
IC: 2513- SERIES CHARACTER GENERATR	Active Electronics	2
IC: 2559 TONE GENERATOR, TELEPHONE	Active Electronics	1
IC: 25XX SERIES CHARACTER GENERATOR	Quest Electronics	6
IC: 2650 SERIES, MICROPROCESSOR	Jameco Electronics	1
IC: 26LS00 SERIES INTERFACE	Active Electronics	4

COMPONENT	COMPANY NAME	VARIETY STOCKED
IC: 26S00 SERIES INTERFACE	Active Electronics	2
IC: 2901 SERIES, SLICE MICROPROC	Active Electronics	12
IC: 2901 SERIES, SLICE MICROPROC	Jameco Electronics	1
IC: 2901 SERIES, SLICE MICROPROC	BCD Radio Parts	4
IC: 2901-4 BIT SLICE MICROPROCESSOR	B.G. Micro	1
IC: 2903-4 BIT SLICE MICROPROCESSOR	B.G. Micro	1
IC: 2911 SEQUENCER	B.G. Micro	1
IC: 29705-16 REGISTER FILES	B.G. Micro	1
IC: 300 SERIES, VOLTAGE REGULATOR	Aldelco	1
IC: 300 SERIES, VOLTAGE REGULATOR	B.G. Micro	2
IC: 300 SERIES, VOLTAGE REGULATOR	Jameco Electronics	5
IC: 3030 DTMF TONE RECEIVER	ITT Microsystems	1
IC: 3031 MF R1 TONE RECEIVER	ITT Microsystems	1
IC: 3032 CCITT #5 TONE RECEIVER	ITT Microsystems	1
IC: 3033 CCITT R2 BACKWARD TONE RCV	ITT Microsystems	1
IC: 3034 CCITT R2 FORWARD TONE RCVR	ITT Microsystems	1
IC: 3040A DTMF LOW GROUP FILTER	ITT Microsystems	1
IC: 3040A DTMF LOW GROUP FILTER	Circuit Specialists	1
IC: 3041A DTMF HIGH GROUP FILTER	ITT Microsystems	1
IC: 3041A DTMF HIGH GROUP FILTER	Circuit Specialists	1
IC: 3044 DTMF LOW GROUP FILTER	ITT Microsystems	1
IC: 3045 DTMF HIGH GROUP FILTER	ITT Microsystems	1
IC: 3064 TRANSMIT PCM FILTER	ITT Microsystems	1
IC: 3065 RECEIVE PCM FILTER	ITT Microsystems	1
IC: 3081/3082 SUBSCRIBER INTERFACE	ITT Microsystems	1
IC: 317 SERIES, VOLTAGE REGULATOR	Aldelco	2
IC: 317 SERIES, VOLTAGE REGULATOR	Jameco Electronics	4
IC: 320 SERIES, VOLTAGE REGULATOR	Electronic Marketplace	2
IC: 320 SERIES, VOLTAGE REGULATOR	Aldelco	17
IC: 320 SERIES, VOLTAGE REGULATOR	Jameco Electronics	15
IC: 3201 DTMFC C-MOS LSI RECEIVER	ITT Microsystems	1
IC: 3210 DTMF TONE RECEIVER	ITT Microsystems	1
IC: 330 SERIES, VOLTAGE REGULATOR	Jameco Electronics	4
IC: 330 SERIES, VOLTAGE REGULATOR	Electronic Marketplace	7
IC: 3300 SERIES FIFO, SHIFT REGISTR	Active Electronics	4
IC: 33000 SERIES SHIFT REGISTER	Active Electronics	2
IC: 3341 FIFO	Fair Radio Sales	1
IC: 3341 FIFO	Daytapro Electronics	1
IC: 3341PC USRT	Quest Electronics	1
IC: 340 SERIES, VOLTAGE REGULATOR	Jameco Electronics	20
IC: 340 SERIES, VOLTAGE REGULATOR	Aldelco	14
IC: 3487 RS 422 LINE DRIVER	Active Electronics	1
IC: 350 SERIES, VOLTAGE REGULATOR	Jameco Electronics	1
IC: 3600 SERIES, INTERFACE	Jameco Electronics	3
IC: 3600 SERIES, KEYBOARD ENCODER	Active Electronics	2
IC: 3820 AM RADIO CHIP	Digital Research	1
IC: 39L00 SERIES LOW CURRENT REG	Circuit Specialists	3
IC: 4000 SERIES, CMOS, DIGITAL	Quest Electronics	64
IC: 4000 SERIES, CMOS, DIGITAL	Babylon Electronics	58
IC: 4000 SERIES, CMOS, DIGITAL	Alpha Electronic Labs	120
IC: 4000 SERIES, CMOS, DIGITAL	Key Electronics	38
IC: 4000 SERIES, CMOS, DIGITAL	Circuit Specialists	48
IC: 4000 SERIES, CMOS, DIGITAL	Aldelco	74
IC: 4000 SERIES, CMOS, DIGITAL	Digi-Key	86
IC: 4000 SERIES, CMOS, DIGITAL	Active Electronics	95
IC: 4000 SERIES, CMOS, DIGITAL	Electronic Marketplace	63
IC: 4000 SERIES, CMOS, DIGITAL	Mouser Electronics	110
IC: 4000 SERIES, CMOS, DIGITAL	B.G. Micro	12
IC: 4000 SERIES, CMOS, DIGITAL	BCD Radio Parts	13

COMPONENT	COMPANY NAME	VARIETY STOCKED
IC: 4000 SERIES, CMOS, DIGITAL	Sintec	42
IC: 4000 SERIES, CMOS, DIGITAL	Semiconductors Surplus	26
IC: 4000 SERIES, CMOS, DIGITAL	Jameco Electronics	95
IC: 4000 SERIES, CMOS, DIGITAL	Proverty Electronics	4
IC: 4194, 4195 TRACKING VOLTAGE REG	Jameco Electronics	2
IC: 5000 SERIES, SHIFT REGISTER	Quest Electronics	2
IC: 5027 CRT CONTROLLER	B.G. Micro	1
IC: 5027 CRT CONTROLLER	BCD Radio Parts	1
IC: 5037 CRT VIDEO TIMER-CONTROLLER	Active Electronics	1
IC: 50380 CLOCK CHIP	BCD Radio Parts	1
IC: 52350 USRT	Quest Electronics	1
IC: 5244 CLOCK DRIVER, CCD	Babylon Electronics	1
IC: 5300 SERIES, CLOCK	Digi-Key	2
IC: 53124 CLOCK CHIP	BCD Radio Parts	1
IC: 5314 DIGITAL CLOCK	Digi-Key	1
IC: 5369 OSC/DIVIDER	Digi-Key	1
IC: 5400 SERIES, TTL, MILITARY SPEC	Babylon Electronics	12
IC: 54L00 SERIES, TTL, MILITARY SPC	Babylon Electronics	3
IC: 5800 SERIES CLOCK/CALENDAR	Digi-Key	3
IC: 5832 CLOCK/CALENDAR	Active Electronics	1
IC: 5837 WHITE NOISE GENERATOR	Digi-Key	1
IC: 5947 LCD DRIVER	Active Electronics	1
IC: 6402 UART	B.G. Micro	1
IC: 6402 UART	BCD Radio Parts	1
IC: 6502 SERIES MICROPROCESSOR	Quest Electronics	10
IC: 6502 SERIES MICROPROCESSOR	Jameco Electronics	1
IC: 6502 SERIES MICROPROCESSOR	Active Electronics	15
IC: 6502 SERIES, COMMODORE	Mouser Electronics	49
IC: 6502 SERIES, ROCKWELL INTRNL	Mouser Electronics	59
IC: 6503 SERIES MICROPROCESSOR	Active Electronics	1
IC: 6504 SERIES MICROPROCESSOR	Active Electronics	1
IC: 6505 SERIES MICROPROCESSOR	Active Electronics	1
IC: 66710 128X9X7 ASCII SHIFT	Mouser Electronics	1
IC: 66740 128X9X7 MATH SYSMBOL ROM	Mouser Electronics	1
IC: 6800 SERIES MICROPROCESSOR	Fuji-Svea	2
IC: 6800 SERIES MICROPROCESSOR	Alpha Electronic Labs	17
IC: 6800 SERIES MICROPROCESSOR	Electronic Marketplace	4
IC: 6800 SERIES MICROPROCESSOR	Circuit Specialists	5
IC: 6800 SERIES MICROPROCESSOR	Quest Electronics	26
IC: 6800 SERIES MICROPROCESSOR	Babylon Electronics	2
IC: 6800 SERIES MICROPROCESSOR	Jameco Electronics	10
IC: 6800 SERIES MICROPROCESSOR	Aldelco	3
IC: 6800 SERIES MICROPROCESSOR	Mouser Electronics	32
IC: 6800 SERIES MICROPROCESSOR	Active Electronics	31
IC: 6800 SERIES MICROPROCESSOR	Semiconductors Surplus	3
IC: 6802 SERIES MICROPROCESSOR	Active Electronics	1
IC: 6802 SERIES MICROPROCESSOR	Jameco Electronics	1
IC: 6808 SERIES MICROPROCESSOR	Active Electronics	1
IC: 6809 SERIES MICROPROCESSOR	Active Electronics	1
IC: 68B00 SERIES MICROPROCESSOR	Quest Electronics	7
IC: 68B21	Semiconductors Surplus	1
IC: 68B45 CRT CONTROLLER	B.G. Micro	1
IC: 68B45 CRT CONTROLLER	BCD Radio Parts	1
IC: 7000 SERIES, TIMER, INTERSIL	Jameco Electronics	1
IC: 7100 SERIES, A/D CONV	Active Electronics	5
IC: 7100 SERIES, A/D CONV	Aldelco	2
IC: 7100 SERIES, A/D CONV, INTERSIL	Jameco Electronics	6
IC: 7200 SERIES	Quest Electronics	4
IC: 7200 SERIES, INTERSIL	Jameco Electronics	23

COMPONENT	COMPANY NAME	VARIETY STOCKED
IC: 7201 SERIAL CONTROLLER	Active Electronics	1
IC: 7205	Electronic Marketplace	1
IC: 7206A, TOUCHTONE ENCODER	Circuit Specialists	1
IC: 723 SERIES, VOLTAGE REGULATOR	Jameco Electronics	2
IC: 7400 SERIES, TTL	B.G. Micro	15
IC: 7400 SERIES, TTL	Key Electronics	15
IC: 7400 SERIES, TTL	Semiconductors Surplus	110
IC: 7400 SERIES, TTL	Active Electronics	218
IC: 7400 SERIES, TTL	Digi-Key	119
IC: 7400 SERIES, TTL	Jameco Electronics	140
IC: 7400 SERIES, TTL	Babylon Electronics	81
IC: 7400 SERIES, TTL	Daytapro Electronics	27
IC: 7400 SERIES, TTL	Sintec	30
IC: 7400 SERIES, TTL	Aldelco	122
IC: 7400 SERIES, TTL	Fuji-Svea	100
IC: 7400 SERIES, TTL	BCD Radio Parts	50
IC: 7400 SERIES, TTL	Circuit Specialists	60
IC: 7400 SERIES, TTL	Quest Electronics	184
IC: 7400 SERIES, TTL	Fair Radio Sales	3
IC: 7400 SERIES, TTL	Electronic Marketplace	99
IC: 7400 SERIES, TTL	Alpha Electronic Labs	136
IC: 74C00 SERIES, CMOS, DIGITAL	Alpha Electronic Labs	54
IC: 74C00 SERIES, CMOS, DIGITAL	Babylon Electronics	11
IC: 74C00 SERIES, CMOS, DIGITAL	Digi-Key	75
IC: 74C00 SERIES, CMOS, DIGITAL	Aldelco	1
IC: 74C00 SERIES, CMOS, DIGITAL	B.G. Micro	1
IC: 74C00 SERIES, CMOS, DIGITAL	Active Electronics	23
IC: 74C00 SERIES, CMOS, DIGITAL	Electronic Marketplace	19
IC: 74C00 SERIES, CMOS, DIGITAL	Quest Electronics	76
IC: 74C00 SERIES, CMOS, DIGITAL	Jameco Electronics	54
IC: 74C9XX, KEYBOARD ENCODER	Circuit Specialists	2
IC: 74C9XX, KEYBOARD ENCODER	Jameco Electronics	2
IC: 74C9XX, KEYBOARD ENCODER	Quest Electronics	2
IC: 74H00 SERIES, TTL	Babylon Electronics	15
IC: 74H00 SERIES, TTL	Jameco Electronics	34
IC: 74HC00 SERIES, TTL	Active Electronics	50
IC: 74L00 SERIES, TTL	Jameco Electronics	33
IC: 74L00 SERIES, TTL	Babylon Electronics	16
IC: 74L00 SERIES, TTL	Quest Electronics	41
IC: 74L00 SERIES, TTL	Sintec	76
IC: 74L00 SERIES, TTL	Electronic Marketplace	18
IC: 74LS00 SERIES, TTL	Jameco Electronics	132
IC: 74LS00 SERIES, TTL	Babylon Electronics	35
IC: 74LS00 SERIES, TTL	Alpha Electronic Labs	138
IC: 74LS00 SERIES, TTL	Aldelco	108
IC: 74LS00 SERIES, TTL	Fuji-Svea	88
IC: 74LS00 SERIES, TTL	Quest Electronics	169
IC: 74LS00 SERIES, TTL	B.G. Micro	54
IC: 74LS00 SERIES, TTL	BCD Radio Parts	56
IC: 74LS00 SERIES, TTL	Digi-Key	115
IC: 74LS00 SERIES, TTL	Circuit Specialists	185
IC: 74LS00 SERIES, TTL	Semiconductors Surplus	86
IC: 74LS00 SERIES, TTL, MOTOROLA	Mouser Electronics	138
IC: 74LS00 SERIES, TTL, SGS	Mouser Electronics	72
IC: 74S00 SERIES, TTL	Quest Electronics	88
IC: 74S00 SERIES, TTL	Electronic Marketplace	28
IC: 74S00 SERIES, TTL	Alpha Electronic Labs	56
IC: 74S00 SERIES, TTL	Semiconductors Surplus	51
IC: 74S00 SERIES, TTL	B.G. Micro	4

COMPONENT	COMPANY NAME	VARIETY STOCKED
IC: 74S00 SERIES, TTL	BCD Radio Parts	5
IC: 74S00 SERIES, TTL	Digi-Key	77
IC: 74S00 SERIES, TTL	Jameco Electronics	85
IC: 74S00 SERIES, TTL	Babylon Electronics	12
IC: 74S00 SERIES, TTL	Sintec	22
IC: 7500 SERIES INTERFACE	Active Electronics	39
IC: 7500 SERIES, MEMORY SENSE AMP	Babylon Electronics	2
IC: 75000 SERIES	Quest Electronics	13
IC: 75000 SERIES	Semiconductors Surplus	17
IC: 75000 SERIES	Electronic Marketplace	6
IC: 75000 SERIES, CMOS TIMER	Jameco Electronics	2
IC: 75000 SERIES, INTERFACE	Jameco Electronics	39
IC: 75000 SERIES, INTERFACE	Aldelco	8
IC: 75000 SERIES, INTERFACE	Digi-Key	9
IC: 75000 SERIES, LED DRIVER	Babylon Electronics	2
IC: 75453 LINE DRIVER	Daytapro Electronics	1
IC: 7600 SERIES, CMOS, INTERSIL	Jameco Electronics	7
IC: 76477, SOUND GENERATOR	Aldelco	1
IC: 76477, SOUND GENERATOR	Electronic Marketplace	1
IC: 76477, SOUND GENERATOR	Key Electronics	1
IC: 76477, SOUND GENERATOR	Jameco Electronics	1
IC: 76477,78,79 SOUND GENERATOR	Active Electronics	3
IC: 76477N, SOUND GENERATOR	Circuit Specialists	1
IC: 76489, SOUND GENERATOR	Circuit Specialists	1
IC: 765 FLOPPY DISK CONTROLLER	Active Electronics	1
IC: 7800 SERIES VOLTAGE REGULATOR	B.G. Micro	4
IC: 7800 SERIES VOLTAGE REGULATOR	Digital Research	4
IC: 7800 SERIES VOLTAGE REGULATOR	BCD Radio Parts	6
IC: 7800 SERIES VOLTAGE REGULATOR	Circuit Specialists	7
IC: 7800 SERIES VOLTAGE REGULATOR	Mouser Electronics	12
IC: 7800 SERIES VOLTAGE REGULATOR	Alpha Electronic Labs	14
IC: 7800 SERIES VOLTAGE REGULATOR	Quest Electronics	10
IC: 7800 SERIES VOLTAGE REGULATOR	Active Electronics	2
IC: 7800 SERIES VOLTAGE REGULATOR	Fuji-Svea	4
IC: 7800 SERIES VOLTAGE REGULATOR	Radiokit	1
IC: 7800 SERIES VOLTAGE REGULATOR	Babylon Electronics	4
IC: 7800 SERIES VOLTAGE REGULATOR	Electronic Marketplace	9
IC: 78G00 SERIES ADJUSTABLE REG	Circuit Specialists	2
IC: 78GK ADJUSTABLE VOLTAGE REG	Babylon Electronics	1
IC: 78GUI VOLTAGE REGULATOR	Alpha Electronic Labs	1
IC: 78H00 SERIES VOLTAGE REGULATOR	Active Electronics	3
IC: 78H00 SERIES VOLTAGE REGULATOR	Alpha Electronic Labs	2
IC: 78H05 VOLTAGE REGULATOR	Quest Electronics	1
IC: 78HG ADJUSTABLE VOLTAGE REG	Active Electronics	1
IC: 78L00 SERIES LOW CURRENT REG	Aldelco	2
IC: 78L00 SERIES LOW CURRENT REG	Babylon Electronics	5
IC: 78L00 SERIES LOW CURRENT REG	Alpha Electronic Labs	4
IC: 78L00 SERIES LOW CURRENT REG	Quest Electronics	4
IC: 78L00 SERIES LOW CURRENT REG	Circuit Specialists	3
IC: 78L00 SERIES LOW CURRENT REG	Electronic Marketplace	3
IC: 78L00 SERIES LOW CURRENT REG	Fuji-Svea	7
IC: 78M00 SERIES VOLTAGE REGULATOR	Fuji-Svea	3
IC: 78M00 SERIES VOLTAGE REGULATOR	Quest Electronics	2
IC: 78M00 SERIES VOLTAGE REGULATOR	Aldelco	1
IC: 78M00 SERIES VOLTAGE REGULATOR	Active Electronics	1
IC: 78M00 SERIES VOLTAGE REGULATOR	Babylon Electronics	2
IC: 78M25 VOLTAGE REGULATOR	Electronic Marketplace	1
IC: 78MG ADJUSTABLE VOLTAGE REG	Babylon Electronics	1
IC: 78MGUI VOLTAGE REGULATOR	Jameco Electronics	1

COMPONENT	COMPANY NAME	VARIETY STOCKED
IC: 78M00 SERIES VOLTAGE REGULATOR	BCD Radio Parts	1
IC: 78P05 5 VOLT, 10 AMP REGULATOR	Active Electronics	1
IC: 78P05 VOLTAGE REGULATOR	Alpha Electronic Labs	1
IC: 78S00 SERIES VOLTAGE REGULATOR	Mouser Electronics	12
IC: 78S40 SWITCHING VOLTAGE REG	Alpha Electronic Labs	1
IC: 7900 SERIES VOLTAGE REGULATOR	Alpha Electronic Labs	13
IC: 7900 SERIES VOLTAGE REGULATOR	Circuit Specialists	7
IC: 7900 SERIES VOLTAGE REGULATOR	Electronic Marketplace	8
IC: 7900 SERIES VOLTAGE REGULATOR	BCD Radio Parts	6
IC: 7900 SERIES VOLTAGE REGULATOR	B.G. Micro	4
IC: 7900 SERIES VOLTAGE REGULATOR	Fuji-Svea	4
IC: 7900 SERIES VOLTAGE REGULATOR	Babylon Electronics	6
IC: 7900 SERIES VOLTAGE REGULATOR	Quest Electronics	9
IC: 79G00 SERIES ADJUSTABLE REG	Circuit Specialists	2
IC: 79L00 SERIES VOLTAGE REGULATOR	Fuji-Svea	4
IC: 79L00 SERIES VOLTAGE REGULATOR	Alpha Electronic Labs	2
IC: 79L00 SERIES VOLTAGE REGULATOR	Quest Electronics	3
IC: 79M00 SERIES VOLTAGE REGULATOR	Active Electronics	3
IC: 79MG ADJUSTABLE VOLTAGE REG	Babylon Electronics	1
IC: 800 SERIES MICROPROCESSOR	Digi-Key	3
IC: 800 SERIES, DTL	Quest Electronics	15
IC: 8000 SERIES, GENERATOR, INTERSIL	Jameco Electronics	3
IC: 8000 SERIES, MOS TELECOM	Mouser Electronics	5
IC: 8000 SERIES, TTL	Jameco Electronics	132
IC: 8000 SERIES, TTL	Semiconductors Surplus	6
IC: 8000 SERIES, TTL	Aldelco	8
IC: 8000 SERIES, TTL	Quest Electronics	99
IC: 8000 SERIES, TTL	Babylon Electronics	48
IC: 8000 SERIES, TTL	BCD Radio Parts	4
IC: 8008-1 MICROPROCESSOR	Fair Radio Sales	1
IC: 8035 SERIES MICROPROCESSOR	Digi-Key	2
IC: 8035 SERIES MICROPROCESSOR	B.G. Micro	1
IC: 8035 SERIES MICROPROCESSOR	BCD Radio Parts	1
IC: 8035 SERIES MICROPROCESSOR	Jameco Electronics	1
IC: 8035 SERIES MICROPROCESSOR	Active Electronics	2
IC: 8038 FUNCTION GENERATOR	Alpha Electronic Labs	1
IC: 8038 FUNCTION GENERATOR	Key Electronics	1
IC: 8038 FUNCTION GENERATOR	Electronic Marketplace	1
IC: 8038 FUNCTION GENERATOR	Circuit Specialists	1
IC: 8038 FUNCTION GENERATOR	Babylon Electronics	1
IC: 8038 FUNCTION GENERATOR	Circuit Specialists	1
IC: 8039 SERIES MICROPROCESSOR	Jameco Electronics	1
IC: 8039 SERIES MICROPROCESSOR	Digi-Key	2
IC: 8039 SERIES MICROPROCESSOR	BCD Radio Parts	1
IC: 8039 SERIES MICROPROCESSOR	B.G. Micro	1
IC: 8039 SERIES MICROPROCESSOR	Active Electronics	2
IC: 8040 SERIES MICROPROCESSOR	Jameco Electronics	1
IC: 8044 CMOS KEYER, CURTIS	Curtis Electro Devices	1
IC: 8044 CMOS KEYER, CURTIS	Alpha Electronic Labs	1
IC: 8060 MICROPROCESSOR	Digi-Key	1
IC: 8070 SERIES MICROPROCESSOR	Jameco Electronics	1
IC: 8073 SERIES MICROPROCESSOR	Digi-Key	1
IC: 8073 SERIES MICROPROCESSOR	Jameco Electronics	1
IC: 8080 SERIES MICROPROCESSOR	Electronic Marketplace	1
IC: 8080 SERIES MICROPROCESSOR	Circuit Specialists	6
IC: 8080 SERIES MICROPROCESSOR	Semiconductors Surplus	2
IC: 8080 SERIES MICROPROCESSOR	Alpha Electronic Labs	4
IC: 8080A SERIES MICROPROCESSOR	Quest Electronics	32
IC: 8080A SERIES MICROPROCESSOR	Active Electronics	38

COMPONENT	COMPANY NAME	VARIETY STOCKED
IC: 8080A SERIES MICROPROCESSOR	B.G. Micro	4
IC: 8080A SERIES MICROPROCESSOR	Jameco Electronics	29
IC: 8080A SERIES MICROPROCESSOR	Digi-Key	1
IC: 8080A SERIES MICROPROCESSOR	Aldelco	1
IC: 8080A SERIES MICROPROCESSOR	Fair Radio Sales	7
IC: 8080A SERIES MICROPROCESSOR	BCD Radio Parts	8
IC: 8085 SERIES MICROPROCESSOR	Electronic Marketplace	1
IC: 8085 SERIES MICROPROCESSOR	Active Electronics	1
IC: 8085 SERIES MICROPROCESSOR	Jameco Electronics	1
IC: 8085 SERIES MICROPROCESSOR	Quest Electronics	1
IC: 8085 SERIES MICROPROCESSOR	BCD Radio Parts	1
IC: 8085 SERIES MICROPROCESSOR	Fair Radio Sales	1
IC: 8086 SERIES MICROPROCESSOR	Active Electronics	1
IC: 8088 SERIES MICROPROCESSOR	Active Electronics	1
IC: 80C00 SERIES, TTL	Active Electronics	2
IC: 80C00 SERIES, TTL	Quest Electronics	4
IC: 80L00 SERIES, TTL	Quest Electronics	11
IC: 80S00 SERIES, TTL	Babylon Electronics	3
IC: 8100 SERIES, NMOS MEMORY	Mouser Electronics	8
IC: 8131 BUS COMPARATOR	B.G. Micro	1
IC: 8131 COMPARATOR	BCD Radio Parts	1
IC: 8200 SERIES	Electronic Marketplace	11
IC: 8200 SERIES	Alpha Electronic Labs	4
IC: 8200 SERIES	Key Electronics	2
IC: 8200 SERIES	Jameco Electronics	34
IC: 8200 SERIES MICROPROCESSORS	Digi-Key	10
IC: 8200 SERIES, VOLT REF, INTERSIL	Jameco Electronics	2
IC: 82S000 SERIES, TTL	Jameco Electronics	25
IC: 8300 SERIES MICROPROCESSORS	Digi-Key	7
IC: 8300 SERIES, INTERFACE	Jameco Electronics	1
IC: 8400 SERIES	Jameco Electronics	7
IC: 8600 SERIES, INTERFACE	Jameco Electronics	4
IC: 8748 SERIES MICROPROCESSOR	BCD Radio Parts	1
IC: 8748 SERIES MICROPROCESSOR	B.G. Micro	1
IC: 8748 SERIES MICROPROCESSOR	Active Electronics	1
IC: 87XX SERIES, A/D CONVERTER	Active Electronics	7
IC: 87XX SERIES, A/D CONVERTER	Quest Electronics	2
IC: 8800 SERIES, INTERFACE	Jameco Electronics	4
IC: 88205-5NC DTMF TONE RECEIVER	ITT Microsystems	1
IC: 8880 PLASMA DISPLAY DRIVER	Babylon Electronics	1
IC: 8H00 SERIES, LOGIC, SIGNETICS	Babylon Electronics	5
IC: 8H000 SERIES, INTERFACE	Quest Electronics	2
IC: 8H000 SERIES, INTERFACE	Jameco Electronics	5
IC: 8T00 SERIES, INTERFACE	Electronic Marketplace	4
IC: 8T00 SERIES, INTERFACE	Babylon Electronics	2
IC: 8T00 SERIES, INTERFACE	Jameco Electronics	2
IC: 8T00 SERIES, INTERFACE	Active Electronics	2
IC: 8T000 SERIES, INTERFACE	Jameco Electronics	18
IC: 8T000 SERIES, INTERFACE	Quest Electronics	25
IC: 900 SERIES, RTL	Babylon Electronics	4
IC: 9000 SERIES, TTL	Electronic Marketplace	6
IC: 9000 SERIES, TTL	BCD Radio Parts	3
IC: 9000 SERIES, TTL	Quest Electronics	36
IC: 9000 SERIES, TTL	Aldelco	7
IC: 9000 SERIES, TTL	Jameco Electronics	32
IC: 9016 HEX DRIVER	Babylon Electronics	1
IC: 90H00 SERIES, TTL	Quest Electronics	1
IC: 9300 SERIES, HEX TO 7-SEGMENT	Alpha Electronic Labs	2
IC: 9300 SERIES, TTL	Active Electronics	27

COMPONENT	COMPANY NAME	VARIETY STOCKED
IC: 9400, A/D CONVERTER	Quest Electronics	1
IC: 9401 CRC GENERATOR/CHECKER	Active Electronics	1
IC: 9403 FIFO	Active Electronics	1
IC: 9528	Circuit Specialists	2
IC: 9582	Circuit Specialists	1
IC: 95H90, PRESCALER, 350 MHZ	Electronic Marketplace	1
IC: 95H90, PRESCALER, 350 MHZ	Aldelco	1
IC: 95H90, PRESCALER, 350 MHZ	Jameco Electronics	1
IC: 95H90, PRESCALER, 350 MHZ, KIT	Circuit Specialists	1
IC: 95H90DCQM, PRESCALER, 350 MHZ	Semiconductors Surplus	1
IC: 9900 SERIES MICROPROCESSOR	Active Electronics	19
IC: 9900 SERIES MICROPROCESSOR	Jameco Electronics	1
IC: 9980 SERIES MICROPROCESSOR	Active Electronics	1
IC: 9981 SERIES MICROPROCESSOR	Active Electronics	1
IC: 9995 SERIES MICROPROCESSOR	Active Electronics	1
IC: AD561 A/D CONVERTER	BCD Radio Parts	1
IC: AD561J D/A CONVERTER	B.G. Micro	1
IC: AD580, VOLTAGE REFERENCE, 2.5 V	Semiconductors Surplus	1
IC: ADC- SERIES, A/D CONVERTER	Digi-Key	12
IC: ADC0800 A/D CONVERTER	Electronic Marketplace	1
IC: ADC0800 SERIES, A/D CONVERTER	Jameco Electronics	3
IC: ADC0804LCN, 9LCN, A/D CONVERTER	Circuit Specialists	2
IC: ADC0809, A/D CONVERTER	Quest Electronics	1
IC: ADD- SERIES, A/D CONVERTER	Digi-Key	2
IC: AF100 SERIES, TOUCH TONE FILTER	Jameco Electronics	3
IC: AF121-1CJ LOW BAND FILTER	Circuit Specialists	1
IC: AF122-1CJ HIGH BAND FILTER	Circuit Specialists	1
IC: AN- SERIES, MATSUSHITA	Fuji-Svea	33
IC: AY3- SERIES, SOUND GENERATOR	Active Electronics	3
IC: AY3-1015 UART	Quest Electronics	1
IC: AY3-1015 UART	Active Electronics	1
IC: AY3-8500-1, TV GAME CHIP	Jameco Electronics	1
IC: AY3-8910 SOUND CHIP	B.G. Micro	1
IC: AY3-8910 SOUND GENERATOR	BCD Radio Parts	1
IC: AY5- SERIES, CLOCK GENERATOR	Jameco Electronics	1
IC: AY5- SERIES, KEYBOARD ENCODER	Aldelco	1
IC: AY5- SERIES, KEYBOARD ENCODER	Jameco Electronics	1
IC: AY5- SERIES, KEYBOARD ENCODER	Quest Electronics	5
IC: AY5- SERIES, TELEPHONE DIALER	Jameco Electronics	2
IC: AY5- SERIES, UART	Quest Electronics	2
IC: AY5-1012 UART	Babylon Electronics	1
IC: AY5-1013 UART	Aldelco	1
IC: AY5-1013 UART	Fair Radio Sales	1
IC: AY5-1013 UART	Active Electronics	1
IC: AY5-1013 UART	Daytapro Electronics	1
IC: AY5-1013 UART	Jameco Electronics	1
IC: BA- SERIES, TOYO, DENGU	Fuji-Svea	2
IC: CA 3XXX SERIES, RCA TYPE	Quest Electronics	35
IC: CA 3XXX SERIES, RCA TYPE	Mouser Electronics	1
IC: CA 3XXX SERIES, RCA TYPE	BCD Radio Parts	5
IC: CA 3XXX SERIES, RCA TYPE	Key Electronics	3
IC: CA 3XXX SERIES, RCA TYPE	Aldelco	2
IC: CA 3XXX SERIES, RCA TYPE	Alpha Electronic Labs	43
IC: CA 3XXX SERIES, RCA TYPE	Semiconductors Surplus	7
IC: CA 3XXX SERIES, RCA TYPE	Jameco Electronics	20
IC: CA 3XXX SERIES, RCA TYPE	Circuit Specialists	37
IC: CA 3XXX SERIES, RCA TYPE	Babylon Electronics	5
IC: CA XXXX SERIES, RCA TYPE	Sintec	34
IC: CALCULATOR CHIP	Jameco Electronics	3

COMPONENT	COMPANY NAME	VARIETY STOCKED
IC: CD- SERIES, CMOS	See IC: 4000 Series	
IC: CLOCK CHIP	Jameco Electronics	15
IC: COM- SERIES UART	Active Electronics	3
IC: COM2017 UART	Quest Electronics	1
IC: COP- SERIES, FLOURESCENT DRIVER	Jameco Electronics	1
IC: COP- SERIES, MICROCONTROLLER	Jameco Electronics	2
IC: COP- SERIES, MICROPROCESSORS	Digi-Key	8
IC: CR- SERIES, REGULATOR	Active Electronics	5
IC: CT- SERIES, CLOCK CHIP	Quest Electronics	2
IC: D 139 SWITCH DRIVER	Active Electronics	1
IC: DAC- SERIES, D/A CONVERTER	Digi-Key	7
IC: DAC- SERIES, D/A CONVERTER	Jameco Electronics	6
IC: DAC- SERIES, D/A CONVERTER	Active Electronics	4
IC: DG- SERIES, ANALOG SWITCH	Active Electronics	12
IC: DISPLAY DRIVER	Jameco Electronics	44
IC: DM8000 SERIES, TTL	See IC: 8000 Series, TTL	
IC: DS 8629 PRESCALER, 135 MHZ	Aldelco	1
IC: DS- SERIES, CLOCK DRIVER	Jameco Electronics	2
IC: DS- SERIES, CLOCK MODULE	Quest Electronics	10
IC: DT1050, DIGITALKER SPEECH SYNTH	Circuit Specialists	1
IC: EAROM	Active Electronics	2
IC: ECT1DB72 SOLID STATE RELAY	MHZ Electronics	
IC: EPROM	Babylon Electronics	3
IC: EPROM	Digi-Key	4
IC: EPROM	Electronic Marketplace	
IC: EPROM	Alpha Electronic Labs	7
IC: EPROM	B.G. Micro	8
IC: EPROM	Semiconductors Surplus	3
IC: EPROM	Aldelco	6
IC: EPROM	Mouser Electronics	5
IC: EPROM	BCD Radio Parts	8
IC: EPROM	Jameco Electronics	10
IC: EPROM	Active Electronics	15
IC: EPROM	Circuit Specialists	3
IC: F 706BPC AUDIO CHIP	Aldelco	1
IC: F 9368 DRIVER LATCH	Aldelco	1
IC: F8-3850 SERIES MICROPROCESSOR	Active Electronics	7
IC: FC47XX SERIES, BIT RATE GEN	Quest Electronics	2
IC: FRI-502 E02 FIFO	Fair Radio Sales	1
IC: GI AY28500-1, GAME CHIP	Quest Electronics	1
IC: HA 2500 OP AMP, HARRIS	Babylon Electronics	1
IC: HA- SERIES, HITACHI	Fuji-Svea	35
IC: HD0165 KEYBOARD ENCODER	Jameco Electronics	1
IC: HD0165-5, KEYBOARD ENCODER	Quest Electronics	1
IC: ICL 710X SERIES, A/D CONVERTER	Quest Electronics	2
IC: ICL 8038 FUNCTION GENERATOR	Jameco Electronics	1
IC: ICL 8069CCQ, VOLTAGE REF, 1.23	Optoelectronics	1
IC: ICM 7045, STOPWATCH	Digi-Key	1
IC: ICM 7045, STOPWATCH	Circuit Specialists	1
IC: ICM 7201, BATTERY INDICATOR	Digi-Key	1
IC: ICM 7202	Electronic Marketplace	1
IC: ICM 7205, STOPWATCH	Circuit Specialists	1
IC: ICM 7205, STOPWATCH	Digi-Key	1
IC: ICM 7207A, OSCILLATOR CONTROL	Circuit Specialists	1
IC: ICM 7208	Electronic Marketplace	1
IC: ICM 7208, 7-DIGIT FREQ COUNTER	Circuit Specialists	1
IC: ICM 7211, LIQUID CRYSTAL DRIVER	Sintec	1
IC: ICM 7215, STOPWATCH	Digi-Key	2
IC: ICM 7216, FREQ COUNTER, 100 MHZ	Digi-Key	2

COMPONENT	COMPANY NAME	VARIETY STOCKED
IC: ICM 7216D,SINGLE CHIP FREQ CNTR	Circuit Specialists	1
IC: ICM 7217, 4 DIGIT UP/DWN CNTR	Digi-Key	1
IC: ICM 7217A, 4 DIGIT UP/DWN CNTR	Circuit Specialists	1
IC: ICM 7218, LED DRIVER	Digi-Key	1
IC: ICM 7226, FREQ COUNTER, 100 MHZ	Digi-Key	2
IC: ICM 7555, 7556 CMOS TIMER	Digi-Key	2
IC: ICM 7555, TIMER	Circuit Specialists	1
IC: ICM 7555, TIMER	Key Electronics	1
IC: IDM 2909 MICROPROGRAM SEQUENCER	Jameco Electronics	1
IC: IM 6402, UART	Digi-Key	1
IC: INS- SERIES	Jameco Electronics	2
IC: INS8073, NSC BASIC MICROINTERP	Circuit Specialists	1
IC: ISP-8F/352 NIBL FIRMWARE ROMS	Digi-Key	1
IC: LA- SERIES, SANYO	Fuji-Svea	21
IC: LAS- SERIES VOLTAGE REGULATOR	BCD Radio Parts	2
IC: LD 111ACJ 3 1/2 DIGIT A/D SET	Active Electronics	1
IC: LD- SERIES, A/D CONVERTER	Quest Electronics	2
IC: LD- SERIES, A/D PROCESSOR	Active Electronics	2
IC: LD- SERIES, SANYO	Fuji-Svea	4
IC: LD110 SERIES, A/D CONVERTER	Jameco Electronics	2
IC: LF- SERIES, BIFET OP AMP	Fuji-Svea	4
IC: LF- SERIES, LINEAR	Circuit Specialists	4
IC: LF- SERIES, LINEAR	Digi-Key	13
IC: LF- SERIES, LINEAR	Jameco Electronics	5
IC: LF- SERIES, LINEAR	Semiconductors Surplus	2
IC: LF- SERIES, LINEAR	Key Electronics	2
IC: LF356BH, JFET OP AMP	Digital Research	1
IC: LH- SERIES, LINEAR	Jameco Electronics	6
IC: LH0021 OP AMP	Babylon Electronics	1
IC: LH0070, VOLTAGE REF, 10V	Quest Electronics	1
IC: LH0070-2H, VOLTAGE REF, 10 V	Semiconductors Surplus	1
IC: LM 2917 FREQ TO VOLTAGE CONV	Circuit Specialists	1
IC: LM 3914N, 15N, LED BAR DRIVER	Circuit Specialists	2
IC: LM- SERIES, LINEAR	Jameco Electronics	104
IC: LM- SERIES, LINEAR	Proverty Electronics	3
IC: LM- SERIES, LINEAR	Radiokit	4
IC: LM- SERIES, LINEAR	Semiconductors Surplus	45
IC: LM- SERIES, LINEAR	Fuji-Svea	17
IC: LM- SERIES, LINEAR	Daytapro Electronics	3
IC: LM- SERIES, LINEAR	Circuit Specialists	91
IC: LM- SERIES, LINEAR	Key Electronics	6
IC: LM- SERIES, LINEAR	BCD Radio Parts	19
IC: LM- SERIES, LINEAR	Active Electronics	104
IC: LM- SERIES, LINEAR	Electronic Marketplace	20
IC: LM- SERIES, LINEAR	Alpha Electronic Labs	79
IC: LM- SERIES, LINEAR	Aldelco	33
IC: LM- SERIES, LINEAR	Babylon Electronics	24
IC: LM- SERIES, LINEAR	Quest Electronics	137
IC: LM- SERIES, LINEAR, MOTOROLA	Mouser Electronics	34
IC: LM- SERIES, VOLTAGE REGULATOR	BCD Radio Parts	3
IC: LM1889, MODULATOR, VIDEO	Quest Electronics	1
IC: LN- SERIES, LINEAR	Digi-Key	154
IC: LU- SERIES, UTILOGIC, SIGNETICS	Babylon Electronics	15
IC: LWM-6205PF, LCD TIME MODULE	Semiconductors Surplus	1
IC: M5XXXX SERIES, MITSUBISHI	Fuji-Svea	10
IC: MA- SERIES, CLOCK MODULE	Quest Electronics	5
IC: MB- SERIES, FUJITSU	Fuji-Svea	4
IC: MC- SERIES	Circuit Specialists	68
IC: MC- SERIES	Daytapro Electronics	3

COMPONENT	COMPANY NAME	VARIETY STOCKED
IC: MC- SERIES	Alaska Microwave Lab	1
IC: MC- SERIES, DIGITAL, MOTOROLA	Mouser Electronics	30
IC: MC- SERIES, LINEAR, MOTOROLA	Mouser Electronics	107
IC: MC- SERIES, MOTOROLA	Semiconductors Surplus	85
IC: MC10000 SERIES, MECL	Fuji-Svea	22
IC: MC1300 SERIES, MDTL	Quest Electronics	1
IC: MC1300 SERIES, MDTL	Fuji-Svea	8
IC: MC1300 SERIES, MDTL	Alpha Electronic Labs	10
IC: MC1300 SERIES, MDTL	Babylon Electronics	1
IC: MC1400 SERIES, MDTL	Radiokit	1
IC: MC1400 SERIES, MDTL	Fuji-Svea	32
IC: MC1400 SERIES, MDTL	Alpha Electronic Labs	19
IC: MC1400 SERIES, MDTL	Jameco Electronics	3
IC: MC14000 SERIES, CMOS	BCD Radio Parts	5
IC: MC14000 SERIES, CMOS	Alpha Electronic Labs	13
IC: MC14000 SERIES, CMOS	Jameco Electronics	1
IC: MC14000 SERIES, CMOS	Fuji-Svea	135
IC: MC140000 SERIES, CMOS, MOTOROLA	Mouser Electronics	72
IC: MC1408 SERIES, A/D CONVERTER	Jameco Electronics	2
IC: MC1408 SERIES, A/D CONVERTER	BCD Radio Parts	1
IC: MC1408 SERIES, A/D CONVERTER	B.G. Micro	1
IC: MC1411 BAUD GENERATOR	Fair Radio Sales	1
IC: MC1469 VOLTAGE REGULATOR	BCD Radio Parts	1
IC: MC1500 SERIES, AMPLIFIER	Babylon Electronics	2
IC: MC1500 SERIES, LINEAR	Alpha Electronic Labs	5
IC: MC1600 SERIES, MECL III	Alpha Electronic Labs	2
IC: MC1600 SERIES, MECL III	Fuji-Svea	2
IC: MC1700 SERIES, LINEAR	Jameco Electronics	1.
IC: MC1700 SERIES, LINEAR	Fuji-Svea	25
IC: MC1700 SERIES, LINEAR	Alpha Electronic Labs	1
IC: MC1800 SERIES, MDTL	Aldelco	1
IC: MC1800 SERIES, MDTL	Fuji-Svea	3
IC: MC1800 SERIES, MDTL	Jameco Electronics	4
IC: MC3000 SERIES, DIGITAL	Jameco Electronics	2
IC: MC3300 SERIES, LINEAR	Alpha Electronic Labs	6
IC: MC3300 SERIES, LINEAR	Fuji-Svea	5
IC: MC3400 SERIES, LINEAR	Alpha Electronic Labs	2
IC: MC3400 SERIES, LINEAR	Fuji-Svea	5
IC: MC3400 SERIES, LINEAR	Jameco Electronics	2
IC: MC4000 SERIES, LINEAR	Jameco Electronics	2
IC: MC4000 SERIES, MTTL	Alpha Electronic Labs	2
IC: MC4000 SERIES, MTTL	Jameco Electronics	1
IC: MC4000 SERIES, MTTL	Fuji-Svea	2
IC: MC600 SERIES, MHTL	Fuji-Svea	5
IC: MC6800 SERIES MICROPROCESSOR	See IC: 6800 Series	
IC: MC6843 FLOPPY DISK CONTROLLER	Alpha Electronic Labs	1
IC: MC75000 SERIES, LINEAR	Fuji-Svea	6
IC: MC800 SERIES, DTL	Babylon Electronics	13
IC: MC800 SERIES, MDTL	Fuji-Svea	3
IC: MC800 SERIES, MDTL	Jameco Electronics	22
IC: MC946, MRTL	Fuji-Svea	1
IC: MCT 2	Electronic Marketplace	1
IC: MF 10CN DUAL ACTIVE FILTER	Digi-Key	1
IC: MFC 6070 ONE WATT AUDIO AMP	Babylon Electronics	1
IC: MK 5002, 4-DIGIT COUNTER	Circuit Specialists	1
IC: MK 5009, TIME BASE	Circuit Specialists	1
IC: MK 50240, OCTAVE TONE GENERATOR	Circuit Specialists	1
IC: MK 50240, OCTAVE TONE GENERATOR	Jameco Electronics	1
IC: MK 5102(N)-5, TONE RECEIVER	Circuit Specialists	1

COMPONENT	COMPANY NAME	VARIETY STOCKED
IC: ML 8204 TELEPHONE TONE RINGER	ETCO	1
IC: MLM- SERIES, LINEAR	Fuji-Svea	11
IC: MM 550 SWITCH	Babylon Electronics	1
IC: MM- SERIES	Electronic Marketplace	4
IC: MM- SERIES	Jameco Electronics	10
IC: MM- SERIES, CPU CLOCK	Jameco Electronics	2
IC: MM- SERIES, KEYBOARD ENCODER	Jameco Electronics	1
IC: MM- SERIES, SPECIAL PURPOSE	Quest Electronics	22
IC: MM- SERIES, TELEPHONE DIALER	Jameco Electronics	2
IC: MM- SERIES, XMTR/RCVR	Jameco Electronics	1
IC: MMH0026, CLOCK DRIVER	Mouser Electronics	1
IC: MMH0026, CLOCK DRIVER, MOS	Fuji-Svea	1
IC: MNQ3799, QUAD TRANSISTOR	ETCO	1
IC: MWA- SERIES RF-IF AMPLIFIER	Alaska Microwave Lab	4
IC: MWA- SERIES RF-IF AMPLIFIER	Circuit Specialists	4
IC: NE- SERIES, LINEAR	Aldelco	7
IC: NE- SERIES, LINEAR	Jameco Electronics	12
IC: NE- SERIES, LINEAR	Alaska Microwave Lab	4
IC: NE- SERIES, LINEAR	Alpha Electronic Labs	10
IC: NE- SERIES, LINEAR	Fuji-Svea	3
IC: NE- SERIES, LINEAR	Semiconductors Surplus	4
IC: NE- SERIES, LINEAR	Radiokit	1
IC: NE- SERIES, LINEAR	Key Electronics	1
IC: NE- SERIES, LINEAR	Electronic Marketplace	5
IC: NE- SERIES, LINEAR	Babylon Electronics	9
IC: NE- SERIES, LINEAR	Quest Electronics	18
IC: PLL- SERIES, NPC	Fuji-Svea	3
IC: PROM	Daytapro Electronics	2
IC: PROM	Sintec	6
IC: PROM	Digi-Key	10
IC: PROM	Quest Electronics	34
IC: PROM	B.G. Micro	2
IC: PROM	Jameco Electronics	28
IC: PROM	BCD Radio Parts	3
IC: PROM, CUSTOM PROGRAMMED	Active Electronics	20
IC: RAM	Electronic Marketplace	
IC: RAM	Jameco Electronics	41
IC: RAM	Fair Radio Sales	1
IC: RAM	Circuit Specialists	4
IC: RAM, BIPOLAR	Active Electronics	10
IC: RAM, DYNAMIC	BCD Radio Parts	5
IC: RAM, DYNAMIC	Aldelco	2
IC: RAM, DYNAMIC	Mouser Electronics	23
IC: RAM, DYNAMIC	Digi-Key	3
IC: RAM, DYNAMIC	Sintec	1
IC: RAM, DYNAMIC	Alpha Electronic Labs	3
IC: RAM, DYNAMIC	Active Electronics	9
IC: RAM, DYNAMIC	B.G. Micro	5
IC: RAM, DYNAMIC	Quest Electronics	15
IC: RAM, DYNAMIC	Babylon Electronics	2
IC: RAM, DYNAMIC	Key Electronics	1
IC: RAM, STATIC	BCD Radio Parts	7
IC: RAM, STATIC	Sintec	2
IC: RAM, STATIC	Aldelco	4
IC: RAM, STATIC	Digital Research	3
IC: RAM, STATIC	Quest Electronics	18
IC: RAM, STATIC	Mouser Electronics	17
IC: RAM, STATIC	Digi-Key	14
IC: RAM, STATIC	Active Electronics	34

COMPONENT	COMPANY NAME	VARIETY STOCKED
IC: RAM, STATIC	B.G. Micro	8
IC: RAM, STATIC	Babylon Electronics	6
IC: RAM, STATIC	Alpha Electronic Labs	19
IC: RC4100 SERIES, LINEAR	Jameco Electronics	2
IC: RC4100 SERIES, LINEAR	Aldelco	2
IC: RC4558 DUAL OP AMP	Fair Radio Sales	1
IC: RM 4136 QUAD 741	Aldelco	1
IC: ROM, CHARACTER GENERATOR	Jameco Electronics	5
IC: ROM, CHARACTER GENERATOR	Babylon Electronics	3
IC: S- SERIES, SANKEN	Fuji-Svea	2
IC: SC42549 SMOKE DETECTOR	Jameco Electronics	1
IC: SE 555	Proverty Electronics	1
IC: SFF- SERIES, TV-CRT CONTROLLER	Mouser Electronics	2
IC: SG 1501 VOLTAGE REGULATOR	Aldelco	1
IC: SG 3501 VOLTAGE REGULATOR, DUAL	Digital Research	1
IC: SG 3501 VOLTAGE REGULATOR, DUAL	BCD Radio Parts	1
IC: SG 4501 VOLTAGE REGULATOR	Aldelco	1
IC: SHIFT REGISTER	Jameco Electronics	17
IC: SHIFT REGISTER	Babylon Electronics	5
IC: SM- SERIES, SANKEN	Fuji-Svea	2
IC: SN7400 SERIES, TTL	See IC: 7400 Series, TTL	
IC: SP8680B, PRESCALER	Optoelectronics	1
IC: SSM2000 SERIES, MUSIC GENERATOR	Quest Electronics	10
IC: STK- SERIES, SANYO	Fuji-Svea	23
IC: TA- SERIES, TOSHIBA	Fuji-Svea	46
IC: TBA- SERIES, HITACHI	Fuji-Svea	1
IC: TC- SERIES, TOSHIBA	Fuji-Svea	4
IC: TCA440, AM RECEIVER	Semiconductors Surplus	1
IC: TD- SERIES, TOSHIBA	Fuji-Svea	2
IC: TL 489C FIVE STEP LEVEL DETECT	Key Electronics	1
IC: TL 497 SWITCHING REGULATOR	Babylon Electronics	1
IC: TL- SERIES, LINEAR	Aldelco	2
IC: TL- SERIES, LINEAR	BCD Radio Parts	2
IC: TL- SERIES, LINEAR	Jameco Electronics	7
IC: TL- SERIES, LINEAR	Active Electronics	40
IC: TLO 61, OP AMP, BIFET	Semiconductors Surplus	1
IC: TLO 81, OP AMP	Semiconductors Surplus	1
IC: TLO 81, OP AMP	Electronic Marketplace	1
IC: TR1602B UART	Quest Electronics	1
IC: UA- SERIES, LINEAR	Daytapro Electronics	3
IC: UA- SERIES, LINEAR	Babylon Electronics	6
IC: UA- SERIES, LINEAR	Proverty Electronics	1
IC: UHIC- SERIES, UNIDEN	Fuji-Svea	4
IC: ULN 2004	Proverty Electronics	1
IC: ULN- SERIES, DARLINGTON ARRAY	Fuji-Svea	2
IC: ULN2000 SERIES, TRANS DAR ARRAY	Mouser Electronics	4
IC: ULN2000 SERIES, TRANS DAR ARRAY	Active Electronics	3
IC: UPC- SERIES, COMPARATOR, NEC	Mouser Electronics	5
IC: UPC- SERIES, D/A CONVERTER, NEC	Mouser Electronics	4
IC: UPC- SERIES, LINEAR, NEC	Mouser Electronics	22
IC: UPC- SERIES, NEC	Fuji-Svea	30
IC: UPC- SERIES, REGULATOR, NEC	Mouser Electronics	16
IC: UPD- SERIES, NEC	Fuji-Svea	5
IC: VOLTAGE REGULATOR, VARIABLE	MHZ Electronics	10
IC: VOLTAGE REGULATOR, VARIABLE	Alpha Electronic Labs	19
IC: XR- SERIES, EXAR	Active Electronics	13
IC: XR- SERIES, EXAR	Quest Electronics	37
IC: XR- SERIES, EXAR	Jameco Electronics	31
IC: XR2206	Electronic Marketplace	1

COMPONENT	COMPANY NAME	VARIETY STOCKED
IC: Z80 SERIES MICROPROCESSOR	Mouser Electronics	12
IC: Z80 SERIES MICROPROCESSOR	Quest Electronics	22
IC: Z80 SERIES MICROPROCESSOR	Alpha Electronic Labs	8
IC: Z80 SERIES MICROPROCESSOR	Circuit Specialists	3
IC: Z80 SERIES MICROPROCESSOR	Semiconductors Surplus	7
IC: Z80 SERIES MICROPROCESSOR	Jameco Electronics	2
IC: Z80, Z80A SERIES MICROPROCESSOR	BCD Radio Parts	6
IC: Z80, Z80A SERIES MICROPROCESSOR	Active Electronics	16
IC: Z80, Z80A SERIES MICROPROCESSOR	Babylon Electronics	9
IC: Z80, Z80A SERIES MICROPROCESSOR	Electronic Marketplace	4
IC: Z8000 SERIES MICROPROCESSOR	Active Electronics	9
IC: Z80A SERIES MICROPROCESSOR	B.G. Micro	6
IC: ZN414, AM RECEIVER, FERRANTI	Circuit Specialists	1
IC: ZN414, AM RECEIVER, FERRANTI	Semiconductors Surplus	1
IMAGE CONDUIT: ROD, 1/8" DIAMETER	Edmund Scientific	1
INDICATOR LIGHT:	See Also: Lamp	
INDICATOR LIGHT: #53 LAMP & SOCKET	Aldelco	1
INDICATOR LIGHT: AMBER LENS, 28 V	Surplus Electronics	1
INDICATOR LIGHT: DOME, DIFFUSE LENS	Mouser Electronics	5
INDICATOR LIGHT: DOME, FRESNEL LENS	Mouser Electronics	4
INDICATOR LIGHT: DOME, JEWEL LENS	Fair Radio Sales	6
INDICATOR LIGHT: DOME, JEWEL LENS	ETCO	8
INDICATOR LIGHT: LED ASSEMBLY	Active Electronics	4
INDICATOR LIGHT: LENS CAPS	Mouser Electronics	8
INDICATOR LIGHT: MINIATURE, LENS	Mouser Electronics	1
INDICATOR LIGHT: NEON, 110 VOLT	Mouser Electronics	2
INDICATOR LIGHT: NEON, JEWEL LENS	Fair Radio Sales	1
INDICATOR LIGHT: NEON, ULTRA MINI	ETCO	3
INDICATOR LIGHT: NEON, ULTRA MINI	Mouser Electronics	3
INDUCTOR FORM CORE: JW MILLER	Circuit Specialists	8
INDUCTOR FORM CORE: JW MILLER	Mouser Electronics	2
INDUCTOR FORM: .3-200 MHZ, SHEILDED	ETCO	12
INDUCTOR FORM: CERAMIC, CAMBION	Semiconductors Surplus	13
INDUCTOR FORM: CERAMIC, MILLEN	Radiokit	4
INDUCTOR FORM: FERRITE, AXIAL LEADS	Digital Research	1
INDUCTOR FORM: FERRITE, AXIAL LEADS	BCD Radio Parts	1
INDUCTOR FORM: JW MILLER	Radiokit	33
INDUCTOR FORM: JW MILLER	Circuit Specialists	29
INDUCTOR FORM: JW MILLER	Mouser Electronics	4
INDUCTOR FORM: PAPER, SLUG, CAMBION	Semiconductors Surplus	2
INDUCTOR FORM: PLUG IN, MILLEN	Radiokit	3
INDUCTOR FORM: SHIELDED IRON POWDER	Amidon Associates	12
INDUCTOR: .39-12.5 uH FIXED VALUE	Digital Research	3
INDUCTOR: .39-500 uH FIXED VALUE	BCD Radio Parts	6
INDUCTOR: 150 mH, TOROIDAL	Fair Radio Sales	1
INDUCTOR: 20 MHY, AUDIO, 350 WV	Star-Tronics	1
INDUCTOR: 20 MHY, CHOKE, AUDIO	Surplus Electronics	1
INDUCTOR: 88 MHY, TOROIDAL	See Inductor: Toroidal	
INDUCTOR: 88 MHY, TOROIDAL	Typetronics	1
INDUCTOR: 88 MHY, TOROIDAL	Amidon Associates	1
INDUCTOR: 2001 ANT ROD, JW MILLER	Circuit Specialists	1
INDUCTOR: 2002 ANT ROD, JW MILLER	Circuit Specialists	1
INDUCTOR: 2002 ANT ROD, JW MILLER	Radiokit	1
INDUCTOR: 2007 ANT ROD, JW MILLER	Radiokit	1

COMPONENT	COMPANY NAME	VARIETY STOCKED
INDUCTOR: 20A- RF CHOKE, JW MILLER	Circuit Specialists	30
INDUCTOR: 21A- RF CHOKE, JW MILLER	Circuit Specialists	29
INDUCTOR: 23A- SERIES, JW MILLER	Mouser Electronics	14
INDUCTOR: 2881 RF CHOKE, JW MILLER	Radiokit	1
INDUCTOR: 32X- ADJ COIL, JW MILLER	Circuit Specialists	12
INDUCTOR: 40A- ADJ COIL, JW MILLER	Radiokit	23
INDUCTOR: 40A- ADJ COIL, JW MILLER	Circuit Specialists	24
INDUCTOR: 41A- ADJ COIL, JW MILLER	Circuit Specialists	24
INDUCTOR: 41A- ADJ COIL, JW MILLER	Radiokit	23
INDUCTOR: 42A- ADJ COIL, JW MILLER	Radiokit	30
INDUCTOR: 42A- ADJ COIL, JW MILLER	Circuit Specialists	32
INDUCTOR: 42XX RF CHOKE, JW MILLER	Radiokit	10
INDUCTOR: 42XX RF CHOKE, JW MILLER	Mouser Electronics	10
INDUCTOR: 42XX RF CHOKE, JW MILLER	Circuit Specialists	6
INDUCTOR: 43A- ADJ COIL, JW MILLER	Circuit Specialists	32
INDUCTOR: 43A- ADJ COIL, JW MILLER	Radiokit	30
INDUCTOR: 43AL,43AV, ANTENNA COILS	Mouser Electronics	2
INDUCTOR: 43LH- RF CHOKE, MOUSER EL	Mouser Electronics	20
INDUCTOR: 43LJ- RF CHOKE, MOUSER EL	Mouser Electronics	15
INDUCTOR: 43LQ- RF CHOKE, MOUSER EL	Mouser Electronics	23
INDUCTOR: 43LR- RF CHOKE, MOUSER EL	Mouser Electronics	27
INDUCTOR: 43LS- RF CHOKE, MOUSER EL	Mouser Electronics	16
INDUCTOR: 43LW 1mH RF CHOKE, MOUSER	Mouser Electronics	1
INDUCTOR: 43LW 5mH RF CHOKE, MOUSER	Mouser Electronics	1
INDUCTOR: 43XX ADJ COIL, JW MILLER	Radiokit	15
INDUCTOR: 43XX ADJ COIL, JW MILLER	Circuit Specialists	16
INDUCTOR: 44XX ADJ COIL, JW MILLER	Circuit Specialists	13
INDUCTOR: 44XX ADJ COIL, JW MILLER	Radiokit	12
INDUCTOR: 45XX ADJ COIL, JW MILLER	Circuit Specialists	15
INDUCTOR: 45XX ADJ COIL, JW MILLER	Radiokit	14
INDUCTOR: 45XX RF CHOKE, JW MILLER	Radiokit	42
INDUCTOR: 45XX RF CHOKE, JW MILLER	Circuit Specialists	31
INDUCTOR: 46- RF CHOKE, JW MILLER	Mouser Electronics	19
INDUCTOR: 46XX RF CHOKE, JW MILLER	Radiokit	32
INDUCTOR: 46XX RF CHOKE, JW MILLER	Circuit Specialists	24
INDUCTOR: 48A- VHF, JW MILLER	Circuit Specialists	9
INDUCTOR: 48A- VHF, JW MILLER	Mouser Electronics	9
INDUCTOR: 49A- VHF, JW MILLER	Circuit Specialists	10
INDUCTOR: 49A- VHF, JW MILLER	Mouser Electronics	10
INDUCTOR: 54XX- HIQ COIL, JW MILLER	Circuit Specialists	20
INDUCTOR: 63XX RF CHOKE, JW MILLER	Radiokit	5
INDUCTOR: 63XX RF CHOKE, JW MILLER	Mouser Electronics	5
INDUCTOR: 63XX RF CHOKE, JW MILLER	Circuit Specialists	5
INDUCTOR: 67-OSC AUTODYNE,JW MILLER	Circuit Specialists	1
INDUCTOR: 70F- RF CHOKE, JW MILLER	Radiokit	87
INDUCTOR: 70F- RF CHOKE, JW MILLER	Circuit Specialists	68
INDUCTOR: 72F- PI COIL, JW MILLER	Circuit Specialists	25
INDUCTOR: 73- UNIVERSAL, JW MILLER	Circuit Specialists	2
INDUCTOR: 73F- RF CHOKE, JW MILLER	Circuit Specialists	25
INDUCTOR: 73F- RF CHOKE, JW MILLER	Radiokit	25
INDUCTOR: 74F- RF CHOKE, JW MILLER	Radiokit	25
INDUCTOR: 78-LINE FILTER, JW MILLER	Circuit Specialists	15
INDUCTOR: 90XX ADJ COIL, JW MILLER	Mouser Electronics	14
INDUCTOR: 90XX ADJ COIL, JW MILLER	Circuit Specialists	14
INDUCTOR: 9210- RF CHOKE, JW MILLER	Circuit Specialists	189
INDUCTOR: 9220- RF CHOKE, JW MILLER	Circuit Specialists	39
INDUCTOR: 9230- RF CHOKE, JW MILLER	Circuit Specialists	48
INDUCTOR: 9250- RF CHOKE, JW MILLER	Mouser Electronics	23
INDUCTOR: 9250- RF CHOKE, JW MILLER	Circuit Specialists	61

COMPONENT	COMPANY NAME	VARIETY STOCKED
INDUCTOR: 9310- RF CHOKE, JW MILLER	Circuit Specialists	27
INDUCTOR: 9320- RF CHOKE, JW MILLER	Circuit Specialists	25
INDUCTOR: 9330- RF CHOKE, JW MILLER	Circuit Specialists	20
INDUCTOR: 9340- RF CHOKE, JW MILLER	Circuit Specialists	25
INDUCTOR: 9350- RF CHOKE, JW MILLER	Circuit Specialists	22
INDUCTOR: 9360- RF CHOKE, JW MILLER	Circuit Specialists	13
INDUCTOR: 9XX RF CHOKE, JW MILLER	Circuit Specialists	25
INDUCTOR: AIR WOUND AIR DUX	Barker & Williamson	68
INDUCTOR: AIR WOUND AIR DUX, B&W	Radiokit	67
INDUCTOR: AIR WOUND MINIDUCTORS	Barker & Williamson	62
INDUCTOR: AIR WOUND MINIDUCTORS,B&W	Radiokit	62
INDUCTOR: ANTENNA ROD/COIL, BC BAND	ETCO	5
INDUCTOR: AUDIO PHASE SHIFT, CUSTOM	Barker & Williamson	
INDUCTOR: CHOKE, AUTOMOBILE, FILTER	Fuji-Svea	2
INDUCTOR: CHOKE, RF	See Inductor: RF Choke	
INDUCTOR: DUAL CHOKE, POWER SUPPLY	Signal Transformer	7
INDUCTOR: FILAMENT CHOKE KIT	Palomar Engineers	1
INDUCTOR: FILAMENT CHOKE KIT	Amidon Associates	1
INDUCTOR: FILAMENT CHOKE KIT	Radiokit	1
INDUCTOR: FILAMENT CHOKE, CUSTOM	Barker & Williamson	
INDUCTOR: FILAMENT CHOKE, HP	Amp Supply	1
INDUCTOR: FILAMENT CHOKE, HP	Barker & Williamson	2
INDUCTOR: FILAMENT CHOKE, HP	ETCO	1
INDUCTOR: FILAMENT CHOKE, HP, B&W	Radiokit	2
INDUCTOR: FILAMENT CHOKE, JW MILLER	Circuit Specialists	1
INDUCTOR: FILTER CHOKE, 1.5-10 HY	Semiconductors Surplus	5
INDUCTOR: FILTER CHOKE,POWER SUPPLY	Signal Transformer	16
INDUCTOR: FILTER CHOKE,POWER SUPPLY	ETCO	1
INDUCTOR: FILTER CHOKE,POWER SUPPLY	Semiconductors Surplus	1
INDUCTOR: FILTER CHOKE,POWER SUPPLY	Fair Radio Sales	44
INDUCTOR: FILTER REACTOR,.01H @ 12A	Surplus Electronics	1
INDUCTOR: HASH CHOKE, JW MILLER	Mouser Electronics	7
INDUCTOR: HASH CHOKE, JW MILLER	Circuit Specialists	7
INDUCTOR: HRO RECEIVER, NATIONAL	Fair Radio Sales	8
INDUCTOR: IF XFMR	ETCO	31
INDUCTOR: IF XFMR, 10.7 M, GOLDWELL	Fuji-Svea	10
INDUCTOR: IF XFMR, 10.7 M, SENTINEL	Mouser Electronics	10
INDUCTOR: IF XFMR, 10.7 M,JW MILLER	Radiokit	6
INDUCTOR: IF XFMR, 10.7 M,JW MILLER	Circuit Specialists	6
INDUCTOR: IF XFMR, 10.7 MHZ	ETCO	6
INDUCTOR: IF XFMR, 100 K, JW MILLER	Radiokit	3
INDUCTOR: IF XFMR, 1610 K,JW MILLER	Radiokit	1
INDUCTOR: IF XFMR, 1650 K,JW MILLER	Radiokit	2
INDUCTOR: IF XFMR, 1800 K,JW MILLER	Radiokit	1
INDUCTOR: IF XFMR, 24 MHZ	Digital Research	1
INDUCTOR: IF XFMR, 24 MHZ	BCD Radio Parts	1
INDUCTOR: IF XFMR, 27 MHZ	Surplus Electronics	4
INDUCTOR: IF XFMR, 27 MHZ	BCD Radio Parts	3
INDUCTOR: IF XFMR, 27 MHZ	Digital Research	3
INDUCTOR: IF XFMR, 4.3 MHZ	Surplus Electronics	3
INDUCTOR: IF XFMR, 4.3 MHZ	Digital Research	3
INDUCTOR: IF XFMR, 4.3 MHZ	BCD Radio Parts	3
INDUCTOR: IF XFMR, 4.55 MHZ	Surplus Electronics	1
INDUCTOR: IF XFMR, 455 K, GOLDWELL	Fuji-Svea	16
INDUCTOR: IF XFMR, 455 K, JW MILLER	Radiokit	21
INDUCTOR: IF XFMR, 455 K, JW MILLER	Circuit Specialists	14
INDUCTOR: IF XFMR, 455 K, MOUSER	Mouser Electronics	4
INDUCTOR: IF XFMR, 455 K, SENTINEL	Mouser Electronics	11
INDUCTOR: IF XFMR, 455 KHZ	Digital Research	4

COMPONENT	COMPANY NAME	VARIETY STOCKED
INDUCTOR: IF XFMR, 455 KHZ	BCD Radio Parts	4
INDUCTOR: IF XFMR, 455 KHZ	ETCO	10
INDUCTOR: IF XFMR, 9 M, JW MILLER	Radiokit	3
INDUCTOR: IF XFMR, VARIABLE FREQ	Mouser Electronics	10
INDUCTOR: IF XFMR,VAR FREQ,GOLDWELL	Fuji-Svea	5
INDUCTOR: PI DUX, B&W	Radiokit	
INDUCTOR: PI DUX, INDENTED	Barker & Williamson	7
INDUCTOR: PI, BAND SWITCHING, B&W	Radiokit	3
INDUCTOR: PI, BAND SWITCHING, HP	Barker & Williamson	3
INDUCTOR: PLATE CHOKE	Semiconductors Surplus	2
INDUCTOR: PLATE CHOKE KIT	Radiokit	1
INDUCTOR: PLATE CHOKE, CUSTOM	Barker & Williamson	
INDUCTOR: PLATE CHOKE, HP	Amp Supply	3
INDUCTOR: PLATE CHOKE, HP	Barker & Williamson	1
INDUCTOR: PLATE CHOKE, HP, B&W	Radiokit	1
INDUCTOR: POT CORE, FERRITE, 123 mH	Fair Radio Sales	1
INDUCTOR: RC- CHOKE, GOLDWELL	Fuji-Svea	31
INDUCTOR: RCH- CHOKE, GOLDWELL	Fuji-Svea	5
INDUCTOR: RF CHOKE	See Also Part Numbers	
INDUCTOR: RF CHOKE	ETCO	4
INDUCTOR: RF CHOKE	BCD Radio Parts	3
INDUCTOR: RF CHOKE	Semiconductors Surplus	100
INDUCTOR: RF CHOKE, .27-50uH, FIXED	Digital Research	3
INDUCTOR: RF CHOKE, FERROXCUBE	Aldelco	2
INDUCTOR: RF CHOKE, HP, HAMMOND	Daytapro Electronics	15
INDUCTOR: RFC- RF CHOKE, JW MILLER	Mouser Electronics	6
INDUCTOR: RFC- RF CHOKE, JW MILLER	Circuit Specialists	7
INDUCTOR: RFC- RF CHOKE, JW MILLER	Radiokit	9
INDUCTOR: ROTARY, HIGH POWER	Radiokit	3
INDUCTOR: ROTARY, HIGH POWER	Barker & Williamson	3
INDUCTOR: ROTARY, HIGH POWER	Fair Radio Sales	2
INDUCTOR: STROBE TRIGGER, 4 KV	ETCO	1
INDUCTOR: TANK COIL, AMP FINAL, 2KW	Amp Supply	1
INDUCTOR: TOROIDAL, 88 MHY	See Inductor: 88 MHY, Toroid	
INDUCTOR: VARIABLE	See Also Inductor: Rotary	
INDUCTOR: VARIABLE, .9-40 uH	Digital Research	5
INDUCTOR: VARIABLE, .9-40 uH	BCD Radio Parts	5
INDUCTOR: VARIABLE, RIBBON, MOTOR	Fair Radio Sales	1
INDUCTOR: VARIABLE, ROLLER	Fair Radio Sales	2
INSULATOR: ANTENNA BASE,HARD RUBBER	Fair Radio Sales	1
INSULATOR: ANTENNA, AN-SULATOR	Lacue Communications	1
INSULATOR: ANTENNA, CENTER, HI-Q	G & C Communications	1
INSULATOR: ANTENNA, CERAMIC	Lacue Communications	1
INSULATOR: ANTENNA, DOG BONE, PORCN	ETCO	1
INSULATOR: ANTENNA, DOG BONE, PORCN	G & C Communications	1
INSULATOR: ANTENNA, DOG BONE, PORCN	Van Gorden Engineering	1
INSULATOR: ANTENNA, END, HI-Q	G & C Communications	1
INSULATOR: ANTENNA, END, POLYMER	Barker & Williamson	1
INSULATOR: ANTENNA, END, POLYMER	Radiokit	1
INSULATOR: ANTENNA, END-SULATOR	Lacue Communications	1
INSULATOR: ANTENNA, HI-Q	Van Gorden Engineering	1
INSULATOR: ANTENNA, POLYMER	Lacue Communications	1
INSULATOR: ANTENNA, PORCELAIN, SPRD	Fair Radio Sales	2
INSULATOR: CERAMIC PAPER	Edmund Scientific	1
INSULATOR: DIPOLE CENTER	Barker & Williamson	2
INSULATOR: DIPOLE CENTER, B&W	Radiokit	2
INSULATOR: EGG, COMPRESSION,CERAMIC	ETCO	1
INSULATOR: FEED-THRU, 3-8" LONG	Fair Radio Sales	5

COMPONENT	COMPANY NAME	VARIETY STOCKED
INSULATOR: FEED-THRU, 3/4 TO 1"	Radiokit	3
INSULATOR: GUYING, CERAMIC	Ham Hardware Headquarters	
INSULATOR: KNOB, STAND-OFF, NAIL MT	ETCO	1
INSULATOR: LINE SPREADER, LUCITE	ETCO	1
INSULATOR: LINENBOARD, BAR, 17" L	Star-Tronics	1
INSULATOR: LINENBOARD, SHEET	Star-Tronics	1
INSULATOR: PORCELAIN, BRASS ENDS	Fair Radio Sales	2
INSULATOR: SCREW HOLE INSERT	Small Parts	22
INSULATOR: SPACER BLOCK, PLASTIC	ETCO	1
INSULATOR: SPACER, THREADED	ETCO	2
INSULATOR: STAND-OFF	BCD Radio Parts	4
INSULATOR: STAND-OFF & FEED THRU	Fair Radio Sales	1
INSULATOR: STAND-OFF, CERAMIC	Semiconductors Surplus	4
INSULATOR: STAND-OFF, CERAMIC	Radiokit	7
INSULATOR: STAND-OFF, CROWS FOOT	Fair Radio Sales	11
INSULATOR: STAND-OFF, PORCELAIN	Fair Radio Sales	4
INSULATOR: STRAIN, PORCELAIN, GUY	Fair Radio Sales	1
INSULATOR: TRANSISTOR, MICA, TO-220	Star-Tronics	ASST
INSULATOR: TRANSISTOR, MICA, TO-3	Mouser Electronics	1
INSULATOR: TRANSISTOR, POWER	Mouser Electronics	1
INSULATOR: TRANSISTOR, POWER	ETCO	7
INSULATOR: TRANSISTOR, TO-220	Surplus Electronics	1
INSULATOR: WASHER, BAKELIGHT, 9/16"	Digital Research	1
INSULATOR: WASHER, MICA, 11/16" OD	Star-Tronics	1
INSULATOR: WASHER, MICA, 5/8" SQ	Digital Research	1
INSULATOR: WASHER, MICA, 5/8" SQ	BCD Radio Parts	1
ISOLATOR: FERRITE, 1.12-140 GHZ	Lectronic	0
ISOLATOR: FERRITE, 3.7-4.2 GHZ	Alaska Microwave Lab	1
ISOLATOR: FERRITE, 4.3 GHZ, COAXIAL	Star-Tronics	1
ISOLATOR: ICL35A, FERRITE, 1.7 GHz	Fair Radio Sales	1
JOYSTICK: 1/4 WATT, LINEAR TAPER	Mouser Electronics	1
JOYSTICK: 100K OHM POTS	Quest Electronics	1
JOYSTICK: ARCADE, WITH SWITCHES	Fair Radio Sales	1
JOYSTICK: JVC, 2-40K OHM, SWITCHES	Semiconductors Surplus	1
JOYSTICK: MICROMINIATURE, PC MOUNT	Jameco Electronics	2
JOYSTICK: OPEN FRAME, 40 & 46K POTS	Semiconductors Surplus	1
JOYSTICK: PLASTIC CASE, CABLE	Jameco Electronics	1
KEYBOARD: 19 KEY, HEX ENCODED	Quest Electronics	1
KEYBOARD: 56 KEY	Quest Electronics	1
KEYBOARD: 56-88 KEY, ALPHA/NUMERIC	Jameco Electronics	6
KEYBOARD: 71 KEY, WORD PROCESSOR	Fair Radio Sales	1
KEYBOARD: 83 KEY, CORTRON	Fair Radio Sales	1
KEYBOARD: PAD	ETCO	8
KEYBOARD: PAD, 12 KEY, TELEPHONE	ETCO	2
KEYBOARD: PAD, 12 KEY, TELEPHONE	Mouser Electronics	1
KEYBOARD: PAD, 12 KEY, TELEPHONE	ETCO	2
KEYBOARD: PAD, 12 KEY, TELEPHONE	Sintec	1
KEYBOARD: PAD, 15 BUTTON	Fair Radio Sales	2
KEYBOARD: PAD, 16 & 25 KEY	BCD Radio Parts	2
KEYBOARD: PAD, 16 KEY	Surplus Electronics	1
KEYBOARD: PAD, 19 KEY, HEXADECIMAL	Jameco Electronics	1
KEYBOARD: PAD, 20 KEY, CALCULATOR	Sintec	1
KEYBOARD: PAD, 20 KEY, CALCULATOR	Digital Research	1
KEYBOARD: PAD, 25 KEY, CALCULATOR	Jameco Electronics	3
KNOB: INSTRUMENT	Fair Radio Sales	22

COMPONENT	COMPANY NAME	VARIETY STOCKED
KNOB: INSTRUMENT	BCD Radio Parts	8
KNOB: INSTRUMENT	BCD Radio Parts	2
KNOB: INSTRUMENT	Digital Research	2
KNOB: INSTRUMENT	Star-Tronics	5
KNOB: INSTRUMENT	Mouser Electronics	49
KNOB: INSTRUMENT, PLASTIC	Active Electronics	4
KNOB: SPIN, FLUTED	Fair Radio Sales	2
KNOB: THUMBWHEEL POTENTIOMETER	Mouser Electronics	2
KNOB: WITH SCALE, MILLEN	Radiokit	6
LABEL: BATTERY POLARITY	Mouser Electronics	1
LABEL: KEYBOARD LEGEND	Mouser Electronics	4
LABEL: SLIDE SWITCH PLATE, 12 PLACE	ETCO	1
LABEL: SOCKET PIN, WIRE WRAP	Quest Electronics	4
LABEL: SOCKET PIN, WIRE WRAP	Sintec	9
LABEL: SOCKET PIN, WIRE WRAP	Digi-Key	9
LABEL: TERMINAL BLOCK, CINCH	ETCO	2
LABEL: TOGGLE SWITCH PLATE, SINGLE	Active Electronics	2
LABEL: WIRE MARKER	Small Parts	12
LAMP:	See Also Indicator Light:	
LAMP: #14,382,327,480,511 INCAND	Fair Radio Sales	5
LAMP: #2 TO #3965, INCANDESCENT	ETCO	78
LAMP: #47, INCANDESCENT	Fuji-Svea	1
LAMP: #49, INCANDESCENT	Fuji-Svea	1
LAMP: #52, INCANDESCENT, 1V	Star-Tronics	1
LAMP: 12 VDC WITH LEADS	Semiconductors Surplus	1
LAMP: 7.5 V, BAYONET BASE	ETCO	1
LAMP: 8 VDC WITH LEADS	Digital Research	1
LAMP: ARGON, PIG TAIL	ETCO	1
LAMP: ARGON, PIG TAIL	Mouser Electronics	1
LAMP: B-3 1/2 BULB, MINI BAYONET	Fuji-Svea	4
LAMP: B-3 1/2 BULB, MINI FLANGE	Mouser Electronics	4
LAMP: F314BW, INTER MARKET	Semiconductors Surplus	1
LAMP: FUSE TYPE, 6.3, 8, 12 VOLT	Fuji-Svea	3
LAMP: FUSE TYPE, 6.3, 8, 14 VOLT	ETCO	3
LAMP: G-3 1/2 BULB, 2-PIN	Fuji-Svea	1
LAMP: G-3 1/2 BULB, 2-PIN	Mouser Electronics	1
LAMP: G-3 1/2 BULB, MINI BAYONET	Mouser Electronics	2
LAMP: G-3 1/2 BULB, MINIATURE SCREW	Mouser Electronics	4
LAMP: G-4 1/2 BULB, BAYONET BASE	Fuji-Svea	1
LAMP: G-4 1/2 BULB, MINIATURE SCREW	Fuji-Svea	2
LAMP: GE 3147, 6V	BCD Radio Parts	1
LAMP: GRAIN OF WHEAT	ETCO	1
LAMP: GRAIN OF WHEAT	Semiconductors Surplus	1
LAMP: GRAIN OF WHEAT, WITH LEADS	Aldelco	2
LAMP: GRAIN OF WHEAT, WITH LEADS	Fuji-Svea	2
LAMP: INCANDESCENT	Surplus Electronics	2
LAMP: INCANDESCENT	Star-Tronics	2
LAMP: INCANDESCENT, TELEPHONE, 12 V	BCD Radio Parts	1
LAMP: NE- SERIES, NEON	ETCO	10
LAMP: NE-2, 2B, NEON	Mouser Electronics	2
LAMP: NE-2, NEON, MIDGET FLANGE	Star-Tronics	1
LAMP: NE-2H, NEON, WITH RESISTOR	BCD Radio Parts	1
LAMP: NE-48, 51, 991, NEON	Fair Radio Sales	3
LAMP: NE-51, NEON	Mouser Electronics	1
LAMP: NEON	ETCO	3
LAMP: NEON	Daytapro Electronics	3
LAMP: PL-63, 80, 120, BAYONET TYPE	Fuji-Svea	3

COMPONENT	COMPANY NAME	VARIETY STOCKED
LAMP: S-8, TRUCK, AUTO, BAYONET	Mouser Electronics	1
LAMP: S-8, TRUCK, AUTO, BAYONET	Fuji-Svea	2
LAMP: T-1 BULB	Mouser Electronics	3
LAMP: T-1 3/4 BULB, BI-PIN BASE	Mouser Electronics	7
LAMP: T-1 3/4 BULB, MIDGET FLANGE	Mouser Electronics	12
LAMP: T-1 3/4 BULB, MIDGET FLANGE	Fuji-Svea	5
LAMP: T-1 3/4 BULB, MIDGET GROOVE	Mouser Electronics	1
LAMP: T-1 3/4 BULB, MIDGET GROOVED	Fuji-Svea	1
LAMP: T-1 3/4 BULB, MIDGET SCREW	Mouser Electronics	5
LAMP: T-1 3/4 BULB, WIRE LEAD	Mouser Electronics	6
LAMP: T-3 1/2 BULB, BAYONET	Fuji-Svea	2
LAMP: T-3 1/4 BULB, BAYONET	Fuji-Svea	3
LAMP: T-3 1/4 BULB, MINI BAYONET	Fuji-Svea	14
LAMP: T-3 1/4 BULB, MINI BAYONET	Mouser Electronics	12
LAMP: T-3 1/4 BULB, MINIATURE SCREW	Mouser Electronics	6
LAMP: T-3 1/4 BULB, SCREW	Fuji-Svea	2
LAMP: T-3 1/4 BULB, SCREW BASE	Fuji-Svea	1
LAMP: T-3 1/4 BULB, WEDGE BASE	Fuji-Svea	1
LAMP: TL-3 BULB, MINIATURE SCREW	Mouser Electronics	1
LAMP: TL-3 BULB, MINIATURE SCREW	Fuji-Svea	1
LAMP: TUNGSTEN HALOGEN, 120 VAC	Edmund Scientific	2
LASER: HeNe TUBE	MHZ Electronics	1
LED:	BCD Radio Parts	8
LED:	Electronic Marketplace	3
LED:	Key Electronics	2
LED:	Alpha Electronic Labs	9
LED:	Quest Electronics	10
LED:	Aldelco	16
LED:	Babylon Electronics	15
LED:	See Also Display:	
LED:	Digital Research	1
LED: .200" DIA	Daytapro Electronics	4
LED: 3mm DIA	Digi-Key	3
LED: 4mm DIA	Digi-Key	3
LED: 5V, BUILT IN RESISTOR	BCD Radio Parts	1
LED: 5V, BUILT IN RESISTOR	Digital Research	1
LED: 5mm DIA	Digi-Key	8
LED: ARROW	Digi-Key	3
LED: ASSEMBLY, INDICATOR	Mouser Electronics	20
LED: ASSEMBLY, RIGHT ANGLE, GREEN	Surplus Electronics	1
LED: BIPOLAR	Digital Research	1
LED: BIPOLAR	BCD Radio Parts	1
LED: CLEAR DOME, RED LIGHT	ETCO	1
LED: CQX- SERIES	Active Electronics	2
LED: DIFFUSED	Jameco Electronics	18
LED: DOT DISPLAY	Digi-Key	1
LED: DUAL COLOR	ETCO	1
LED: DUAL COLOR	Sintec	3
LED: DUAL COLOR	Jameco Electronics	1
LED: DUAL COLOR	Mouser Electronics	2
LED: DUAL COLOR	Aldelco	2
LED: FLV 5007, FAIRCHILD	Semiconductors Surplus	1
LED: FLV- SERIES	Active Electronics	17
LED: FPA- SERIES	Active Electronics	2
LED: GL- SERIES	Active Electronics	2
LED: HOLDER	Digital Research	1
LED: INFRARED	Jameco Electronics	1

COMPONENT	COMPANY NAME	VARIETY STOCKED
LED: LD- SERIES	Active Electronics	10
LED: LED 209, T-1	Active Electronics	1
LED: MLED SERIES, MOTOROLA	Fuji-Svea	3
LED: MOUNTING SYSTEM, CLIP- SERIES	Active Electronics	2
LED: MOUNTING SYSTEM, CLIPLITE	Key Electronics	5
LED: MOUNTING SYSTEM, CLIPLITE	Daytapro Electronics	4
LED: MOUNTING SYSTEM, CLIPLITE	Quest Electronics	1
LED: MOUNTING SYSTEM, CLIPLITE	Jameco Electronics	5
LED: MOUNTING SYSTEM, CUBELITE	Daytapro Electronics	4
LED: MOUNTING SYSTEM, PANEL	Daytapro Electronics	1
LED: MOUNTING SYSTEM, T-1 3/4	Semiconductors Surplus	1
LED: MOUNTING SYSTEM, T-1 3/4	Mouser Electronics	11
LED: MOUNTING SYSTEM, T-1 3/4	Digi-Key	1
LED: MOUNTING SYSTEM, T-1, T-1 3/4	Sintec	6
LED: MV- SERIES	Active Electronics	28
LED: NDL SERIES, NEC	California Eastern Lab	2
LED: NSL SERIES, T-1 3/4	Digi-Key	1
LED: PIN POINT	Mouser Electronics	1
LED: POINT SOURCE	Jameco Electronics	4
LED: RBG 1000 RED BAR GRAPH	Active Electronics	1
LED: RECTANGULAR	Jameco Electronics	3
LED: RECTANGULAR	Mouser Electronics	13
LED: RECTANGULAR	Digi-Key	3
LED: RL4850 RED T-1 3/4, NO MIN MCD	Active Electronics	1
LED: RLC- SERIES, CONSTANT CURRENT	Active Electronics	3
LED: SQUARE, 5x5mm	Mouser Electronics	1
LED: SUPER BRIGHT, STANLEY	Active Electronics	22
LED: T-0 3/4	Sintec	4
LED: T-0 3/4	Mouser Electronics	1
LED: T-1 MINIATURE	ETCO	3
LED: T-1 MINIATURE	Mouser Electronics	20
LED: T-1 MINIATURE	Sintec	5
LED: T-1 3/4, JUMBO	Digi-Key	2
LED: T-1 3/4, JUMBO	Mouser Electronics	32
LED: T-1 3/4, JUMBO	Sintec	17
LED: T-1 3/4, JUMBO	Semiconductors Surplus	3
LED: T-1 3/4, JUMBO	ETCO	4
LED: T-1 3/4, JUMBO	Fair Radio Sales	2
LED: T-1 3/4, SOLID STATE BI-PIN	Mouser Electronics	1
LED: T-2	Mouser Electronics	16
LED: TIL- SERIES, INFRARED	Active Electronics	4
LED: TIL- SERIES, VISIBLE	Active Electronics	7
LED: TRIANGULAR, .2"	Jameco Electronics	3
LED: TRIANGULAR, 4.3mm	Mouser Electronics	1
LENS MOUNT: VIDICON TYPE C	ATV Research	1
LENS: OPTICAL	Edmund Scientific	544
LENS: VIDEO CAMERA	ATV Research	39
LENS: VIDEO CAMERA	Semiconductors Surplus	2
LIMITER: RF, TO 900 MHZ	Mini-Circuits Laboratory	3
LIMITER: WAVEGUIDE, DIODE	Lectronic	
LINE STRETCHER: COAXIAL	Lectronic	
LOAD: MICROWAVE	Lectronic	
LOAD: MICROWAVE, 50 OHM, .5-2 WATT	Elcom Systems	6
LOAD: MICROWAVE, 75 OHM, .25-1 WATT	Elcom Systems	2

COMPONENT	COMPANY NAME	VARIETY STOCKED
LOAD: MICROWAVE, 93 OHM, .25-1 WATT	Elcom Systems	2
LOAD: RF, COAXIAL, 50 OHM, 2W-80 KW	Fuji-Svea	43
LOAD: RF, COAXIAL, 73 OHM, F FEMALE	Aldelco	1
MAGNET: ELECTRO-MAGNET	Edmund Scientific	2
MAGNET: PERMANENT	Edmund Scientific	83
MAGNET: PERMANENT	ETCO	3
MAGNETIC PICK-UP: VOLTAGE TRANSDUCR	Surplus Electronics	1
MAGNETIC SHIELD: SHEET	ATV Research	3
MATRIX BOARD: PROGRAMMABLE	Sintec	1
METER SHUNT: .0025 OHM, 20 AMP	Star-Tronics	1
METER SHUNT: ADJUSTABLE, HI CURRENT	Sintec	2
METER: "S" UNITS	ETCO	5
METER: "S" UNITS, ILLUMINATED	Surplus Electronics	2
METER: BATTERY INDICATOR	Mouser Electronics	2
METER: BATTERY INDICATOR	Surplus Electronics	1
METER: DIGITAL VOLTMETER, PANEL MT	Fuji-Svea	14
METER: EDGE MOUNT	Mouser Electronics	15
METER: EDGE MOUNT	Daytapro Electronics	8
METER: EDGE MOUNT	Sintec	9
METER: EDGE MOUNT	Star-Tronics	1
METER: EDGE MOUNT, MINIATURE	Circuit Specialists	2
METER: EDGE MOUNT, MINIATURE	Mouser Electronics	2
METER: ILLUMINATION KIT	Mouser Electronics	1
METER: ILLUMINATION KIT	Sintec	1
METER: INDUCTION AMMETER	Sintec	1
METER: PANEL MOUNT	Sintec	60
METER: PANEL MOUNT	Amp Supply	7
METER: PANEL MOUNT	Mouser Electronics	56
METER: PANEL MOUNT	Star-Tronics	7
METER: PANEL MOUNT	ETCO	33
METER: PANEL MOUNT	Aldelco	2
METER: PANEL MOUNT	Semiconductors Surplus	5
METER: PANEL MOUNT	BCD Radio Parts	1
METER: PANEL MOUNT	Surplus Electronics	6
METER: PANEL MOUNT, GIANT 9x8"	ETCO	1
METER: PANEL MOUNT, PRECISION, 2%	ETCO	8
METER: PANEL, ILLUMINATED	Daytapro Electronics	5
METER: PANEL, ILLUMINATED, MOUSER	Mouser Electronics	1
METER: PANEL, MINIATURE	Circuit Specialists	4
METER: PANEL, MINIATURE, MOUSER	Mouser Electronics	3
METER: PANEL, PRECISION	Circuit Specialists	9
METER: PANEL, PRECISION, MOUSER	Mouser Electronics	15
METER: PANEL, SQUARE	Daytapro Electronics	25
METER: SIGNAL STRENGTH, dB	Surplus Electronics	1
METER: VU-TYPE, PANEL MOUNT	Mouser Electronics	2
METER: VU-TYPE, PANEL MOUNT	ETCO	3
METER: VU-TYPE, PANEL MOUNT	Sintec	4
METER: ZERO-CENTER	Surplus Electronics	1
METER: ZERO-CENTER, NO CASE	Star-Tronics	1
MICROPHONE ELEMENT: CARBON	ETCO	1
MICROPHONE ELEMENT: CARBON	Fair Radio Sales	3
MICROPHONE ELEMENT: CONDENSER	ETCO	3

COMPONENT	COMPANY NAME	VARIETY STOCKED
MICROPHONE ELEMENT: CONDENSER	Mouser Electronics	7
MICROPHONE ELEMENT: CRYSTAL	ETCO	3
MICROPHONE ELEMENT: CRYSTAL/CERAMIC	Mouser Electronics	9
MICROPHONE ELEMENT: DYNAMIC	Mouser Electronics	5
MICROPHONE ELEMENT: MUSIC INSTRUMNT	ETCO	18
MICROPHONE TRANSFORMER: MATCHING	ETCO	4
MICROPHONE WINDSCREEN:	ETCO	2
MIRROR: OPTICAL	Edmund Scientific	262
MIXER: .03-4 GHZ, DOUBLE BALANCED	Elcom Systems	1
MIXER: .07-200 & .1-400 MHZ	MHZ Electronics	2
MIXER: .2-300 MHZ, DOUBLE BALANCED	MHZ Electronics	1
MIXER: .9-1.3 GHZ	Alaska Microwave Lab	1
MIXER: 2-500 MHZ, DOUBLE BALANCED	Elcom Systems	1
MIXER: 2-500 MHZ, HEWLETT PACKARD	MHZ Electronics	2
MIXER: 3.7-4.2 GHZ	MHZ Electronics	1
MIXER: 3.7-4.2 GHZ	Alaska Microwave Lab	1
MIXER: 5-1000 MHZ, DOUBLE BALANCED	Elcom Systems	1
MIXER: 500 HZ TO 3 GHZ, DBLE BAL	Mini-Circuits Laboratory	125
MIXER: CRYSTAL, MICROWAVE	Lectronic	
MIXER: HF, TRANSISTOR	International Crystal	1
MIXER: WAVEGUIDE	Lectronic	
MIXER: WR90 WAVEGUIDE	Alaska Microwave Lab	1
MODULATOR: AM, 60-150 MHZ	MHZ Electronics	1
MODULATOR: PXP-4500, VIDEO TO RF	ATV Research	1
MODULATOR: TXA5-4 ATV VIDEO, 435MHz	P.C. Electronics	1
MODULATOR: UM1082, VIDEO, ASTEC	Semiconductors Surplus	1
MOTOR: 115 VAC	Fair Radio Sales	12
MOTOR: 115 VAC	ETCO	2
MOTOR: 115 VAC, GEARED	ETCO	4
MOTOR: 115 VAC, GEARED, 4 RPM	Surplus Electronics	1
MOTOR: 115 VAC, MITE	Atlantic Surplus Sales	1
MOTOR: 24 VAC	ETCO	3
MOTOR: DC	Fair Radio Sales	18
MOTOR: DC, HEAVY DUTY, 1.5-3 VOLT	Digital Research	1
MOTOR: DC, HEAVY DUTY, 1.5-3 VOLT	BCD Radio Parts	1
MOTOR: DC, LV	ETCO	13
MOTOR: DC, SMALL	Edmund Scientific	18
MOTOR: DC, SMALL	Semiconductors Surplus	9
MOTOR: DC, SMALL, 3-10 VOLT	Surplus Electronics	1
MOTOR: DC, SMALL, TOY	Jameco Electronics	1
MOTOR: GEARED, 115 VAC	ETCO	4
MOTOR: RATCHET, 24 V DC	Fair Radio Sales	1
MOTOR: STEPPER	Semiconductors Surplus	1
MOTOR: STEPPER, 12 V DC	Fair Radio Sales	1
MOTOR: TELETYPE	Typetronics	
NOISE SOURCE: MICROWAVE, WAVEGUIDE	Lectronic	
OPTO COUPLER: 4N- SERIES	Fuji-Svea	5
OPTO COUPLER: 4N26, FAIRCHILD	Semiconductors Surplus	1
OPTO COUPLER: MOC- SERIES	Fuji-Svea	2
OPTO COUPLER: PS2000 SERIES, NEC	Mouser Electronics	3
OPTO COUPLER: TIP- SERIES	Fuji-Svea	30

COMPONENT	COMPANY NAME	VARIETY STOCKED
OPTO ISOLATOR:	Jameco Electronics	5
OPTO ISOLATOR:	Circuit Specialists	4
OPTO ISOLATOR:	Quest Electronics	4
OPTO ISOLATOR:	Babylon Electronics	3
OPTO ISOLATOR: 10 KV, OPCOA	Sintec	2
OPTO ISOLATOR: 4N- SERIES	Active Electronics	11
OPTO ISOLATOR: 6N- SERIES	Active Electronics	4
OPTO ISOLATOR: IL- SERIES	Active Electronics	2
OSCILLATOR: 11 GHZ, CRYSTAL CONTR	MHZ Electronics	1
OSCILLATOR: CRYSTAL, CLOCK, TTL, 5V	Jameco Electronics	10
OSCILLATOR: CRYSTAL, HF	International Crystal	3
OSCILLATOR: CRYSTAL, LF	International Crystal	1
OSCILLATOR: GUNN, 10.5 GHZ	Alaska Microwave Lab	1
OSCILLATOR: GUNN, 12-18 GHZ	MHZ Electronics	1
OSCILLATOR: GUNN, 8-12.4 GHZ	MHZ Electronics	1
OSCILLATOR: GUNN, LOCKED, 4.5-6 GHZ	Lectronic	1
OSCILLATOR: HF WITH CRYSTAL	International Crystal	
OSCILLATOR: IMPATT, 10.5 GHZ	Alaska Microwave Lab	1
OSCILLATOR: IMPATT, 10.6 GHZ	Lectronic	1
OSCILLATOR: KLYSTRON, .95-2 GHZ	Lectronic	
OSCILLATOR: KLYSTRON, 10 GHZ	Lectronic	1
OSCILLATOR: KLYSTRON, 7-11 GHZ	Lectronic	
OSCILLATOR: MC6871, 1 & 2 MHZ, MOT	BCD Radio Parts	1
OSCILLATOR: MC6871, 1 & 2 MHZ, MOT	Digital Research	1
OSCILLATOR: VOLTAGE CONT, 13.3 GHZ	Lectronic	1
OSCILLATOR: VOLTAGE CONT, 7-11 GHZ	Lectronic	1
OSCILLATOR: VOLTAGE CONT,3.6-4.2GHZ	MHZ Electronics	1
OSCILLATOR: VOLTAGE TUNED, 2.7 GHZ	Alaska Microwave Lab	1
OSCILLATOR: VOLTAGE TUNED, 3.6 GHZ	Alaska Microwave Lab	1
OVEN: CRYSTAL, 10 DEG C, 6/12 VOLT	ETCO	1
PANEL:	See Sheet Metal:	
PATCH PANEL: 1/4" PLUG, 19" PANEL	Atlantic Surplus Sales	2
PHASE SHIFT NETWORK: AUDIO	Barker & Williamson	1
PHASE SHIFTER: WAVEGUIDE	Lectronic	
PHOTO CELL:	Key Electronics	1
PHOTO CELL: CADMIUM SULPHIDE	Mouser Electronics	1
PHOTO CELL: CADMIUM SULPHIDE, DUAL	Babylon Electronics	1
PHOTO CELL: FPT- SERIES, PHOTO/INFR	Active Electronics	3
PHOTO CELL: SUBMINIATURE	Digital Research	1
PHOTO CELL: SUBMINIATURE, 1-50K OHM	BCD Radio Parts	1
PHOTO COUPLER:	See Opto Coupler:	
POTENTIOMETER:	See Resistor: Variable	
POTTING: CAPS, CRIMPED CONNECTION	Digi-Key	5
POWER DIVIDER:	See Splitter: RF	
PRISM: OPTICAL	Edmund Scientific	59

COMPONENT	COMPANY NAME	VARIETY STOCKED
PULLEY: BEARING MOUNT, NYLON	Small Parts	6
PULLEY: HIGH STRENGTH NYLON, 1-6"OD	Small Parts	19
PULLEY: IDLER, DELRIN	Small Parts	7
PULLEY: SPROCKET, ROLLER CHAIN	Small Parts	12
PULLEY: TIMING, TOOTHED, NYLON	Small Parts	42
R.F. CHOKE:	See Inductor	
RELAY: ANTENNA CHANGE OVER, 2 KW	Amp Supply	2
RELAY: ART-13, VACUUM, KEYING	Fair Radio Sales	1
RELAY: COAX	See Also Switch: RF	
RELAY: COAX, RF	Lectronic	
RELAY: COAX, RF, 100 W, MAGNECRAFT	Semiconductors Surplus	1
RELAY: COAX, RF, 1000 W CW	Barker & Williamson	1
RELAY: COAX, RF, 1000 W CW, B&W	Radiokit	1
RELAY: COAX, RF, AMPHENOL	Semiconductors Surplus	3
RELAY: COAX, RF, TRANSCO	Semiconductors Surplus	6
RELAY: DIP	Alpha Electronic Labs	4
RELAY: DIP, 115 VAC COIL	Jameco Electronics	1
RELAY: DIP, 12V COIL	Circuit Specialists	4
RELAY: DIP, 5 TO 12 VDC COIL	Jameco Electronics	4
RELAY: DIP, 5V COIL	Circuit Specialists	2
RELAY: DIP, REED, 5 TO 24 VDC COIL	Active Electronics	8
RELAY: DIP, REED, 5-12 VOLT	Mouser Electronics	16
RELAY: DUAL COIL, 24 V DC	Fair Radio Sales	1
RELAY: GENERAL PURPOSE	Alpha Electronic Labs	9
RELAY: GENERAL PURPOSE	Digi-Key	16
RELAY: GENERAL PURPOSE	Mouser Electronics	15
RELAY: GENERAL PURPOSE, 12 VDC	Digital Research	1
RELAY: GENERAL PURPOSE, 24 VDC	BCD Radio Parts	1
RELAY: GENERAL PURPOSE, 6-24 VDC	Active Electronics	6
RELAY: GENERAL PURPOSE, MINIATURE	Mouser Electronics	5
RELAY: MINIATURE, PC MT	Circuit Specialists	2
RELAY: OPEN FRAME, 24 VDC	Star-Tronics	1
RELAY: PC BOARD MOUNT, LV COIL	Circuit Specialists	4
RELAY: PC BOARD MOUNT, MINIATURE	Digi-Key	8
RELAY: PLATE	Fair Radio Sales	4
RELAY: POWER, 30 AMP	Mouser Electronics	2
RELAY: POWER, HEAVY CONTACT	Circuit Specialists	4
RELAY: REED	Alpha Electronic Labs	4
RELAY: REED	Jameco Electronics	6
RELAY: REED	ETCO	2
RELAY: REED WITH COIL	Babylon Electronics	2
RELAY: REED WITH COIL	Circuit Specialists	4
RELAY: REED WITH COIL, 5 V	BCD Radio Parts	1
RELAY: REED WITH COIL, 5 V	Digital Research	1
RELAY: REED, DUAL COIL, 3 V	BCD Radio Parts	1
RELAY: REED, DUAL COIL, 3 V	Digital Research	1
RELAY: SENSITIVE COIL	Fair Radio Sales	6
RELAY: SMALL, PRECISION	Circuit Specialists	3
RELAY: SOLID STATE, 120-240 V AC	Semiconductors Surplus	5
RELAY: SOLID STATE, TECCOR	Mouser Electronics	13
RELAY: STANDARD	ETCO	23
RELAY: STANDARD	Key Electronics	4
RELAY: STANDARD	Fair Radio Sales	40
RELAY: STANDARD	Semiconductors Surplus	16
RELAY: STANDARD	Surplus Electronics	12
RELAY: STANDARD, 12 VDC COIL	Star-Tronics	1
RELAY: STANDARD, 24 VDC, 4PDT	Babylon Electronics	1

COMPONENT	COMPANY NAME	VARIETY STOCKED
RELAY: SUBMINI, LV, SENSITIVE COIL	Circuit Specialists	3
RELAY: TIME DELAY	ETCO	2
RELAY: TIME DELAY	Fair Radio Sales	2
RELAY: TIME DELAY, 110 VDC, .2-60 S	Star-Tronics	1
RELAY: TO-5 TYPE	Semiconductors Surplus	1
RESISTOR: 0.1%, 15 OHM, 7 MATCHED	Star-Tronics	1
RESISTOR: 1%, 1K OHM	Digital Research	1
RESISTOR: CARBON COMP	ETCO	17
RESISTOR: CARBON COMP, 1 WATT	Star-Tronics	8
RESISTOR: CARBON COMP, 1/2 WATT	Radiokit	
RESISTOR: CARBON COMP, 1/2 WATT	Active Electronics	184
RESISTOR: CARBON COMP, 1/2 WATT	Mouser Electronics	163
RESISTOR: CARBON COMP, 1/4 WATT	Mouser Electronics	159
RESISTOR: CARBON COMP, 1/4 WATT	Star-Tronics	13
RESISTOR: CARBON COMP, 1/4 WATT	Quest Electronics	FULL
RESISTOR: CARBON COMP, 1/4 WATT	Radiokit	
RESISTOR: CARBON COMP, 1/4 WATT	ETCO	4
RESISTOR: CARBON COMP, 1/4 WATT	Active Electronics	209
RESISTOR: CARBON COMP, 1/4 WATT, AB	Semiconductors Surplus	33
RESISTOR: CARBON FILM, 1 WATT	Proverty Electronics	
RESISTOR: CARBON FILM, 1 WATT	Mouser Electronics	57
RESISTOR: CARBON FILM, 1/2 WATT	Proverty Electronics	
RESISTOR: CARBON FILM, 1/2 WATT	Fuji-Svea	155
RESISTOR: CARBON FILM, 1/2 WATT	Mouser Electronics	145
RESISTOR: CARBON FILM, 1/2 WATT	Circuit Specialists	67
RESISTOR: CARBON FILM, 1/2 WATT	Digi-Key	170
RESISTOR: CARBON FILM, 1/2 WATT	Key Electronics	23
RESISTOR: CARBON FILM, 1/4 WATT	Digi-Key	170
RESISTOR: CARBON FILM, 1/4 WATT	Jameco Electronics	161
RESISTOR: CARBON FILM, 1/4 WATT	Fuji-Svea	155
RESISTOR: CARBON FILM, 1/4 WATT	Mouser Electronics	53
RESISTOR: CARBON FILM, 1/4 WATT	Key Electronics	40
RESISTOR: CARBON FILM, 1/4 WATT	Proverty Electronics	
RESISTOR: CARBON FILM, 1/4 WATT	Babylon Electronics	135
RESISTOR: CARBON FILM, 1/4 WATT	Circuit Specialists	124
RESISTOR: FLAME PROOF, > 1 WATT	Daytapro Electronics	62
RESISTOR: HIGH VOLT, 44 & 66 MEGOHM	ETCO	2
RESISTOR: JOYSTICK	See Joystick	
RESISTOR: METAL FILM, 1%, 1/2 WATT	Mouser Electronics	148
RESISTOR: METAL FILM, 1%, 1/4 WATT	Mouser Electronics	143
RESISTOR: METAL FILM, 1%, 1/4 WATT	Proverty Electronics	
RESISTOR: METAL FILM, 1%, 1/4 WATT	Sintec	3
RESISTOR: METAL FILM, 2%, 1/4 WATT	Proverty Electronics	
RESISTOR: METAL FILM, 5%, 1&2 WATT	Mouser Electronics	39
RESISTOR: METAL FILM, 5%, 1/4 WATT	Proverty Electronics	
RESISTOR: METAL OXIDE, 5%, 1&2 WATT	Mouser Electronics	83
RESISTOR: MICROWAVE CHIP	Alaska Microwave Lab	2
RESISTOR: MVX-3, 1000 MEG OHM, TRW	Semiconductors Surplus	1
RESISTOR: NETWORK, 13 X 220 K OHM	BCD Radio Parts	1
RESISTOR: NETWORK, 13 X 220 K OHM	Digital Research	1
RESISTOR: NETWORK, DIP	Active Electronics	490
RESISTOR: NETWORK, DIP	Alpha Electronic Labs	
RESISTOR: NETWORK, DIP	ETCO	2
RESISTOR: NETWORK, SIP	Sintec	9
RESISTOR: NETWORK, SIP	Digi-Key	75
RESISTOR: NETWORK, SIP	Alpha Electronic Labs	
RESISTOR: POWER	See Also Wirewound	
RESISTOR: POWER	ETCO	23

COMPONENT	COMPANY NAME	VARIETY STOCKED
RESISTOR: POWER	Digital Research	2
RESISTOR: POWER, 10 MEG OHM, 5 WATT	Star-Tronics	1
RESISTOR: POWER, 3 TO 20 WATT	Alpha Electronic Labs	3
RESISTOR: POWER, 3 TO 220 WATT	Fair Radio Sales	23
RESISTOR: POWER, 5 & 10 WATT	Circuit Specialists	20
RESISTOR: POWER, 5 TO 100 WATT	Star-Tronics	2
RESISTOR: VARIABLE, 10 TURN, MINI	Circuit Specialists	5
RESISTOR: VARIABLE, 12 TURN, DIP	Babylon Electronics	1
RESISTOR: VARIABLE, 15 TURN, MINI	Jameco Electronics	13
RESISTOR: VARIABLE, 16 TURN, PC	Circuit Specialists	6
RESISTOR: VARIABLE, 18 TURN, PC	Active Electronics	1
RESISTOR: VARIABLE, 2 LEG	Key Electronics	3
RESISTOR: VARIABLE, 2.75 OHM, 200W	Fair Radio Sales	1
RESISTOR: VARIABLE, 20 TURN, 1 W	ETCO	1
RESISTOR: VARIABLE, 20 TURN, TRIM	Quest Electronics	19
RESISTOR: VARIABLE, 25 TURN W/KNOB	Key Electronics	1
RESISTOR: VARIABLE, 35 TURN, TRIM	Star-Tronics	1
RESISTOR: VARIABLE, 9 TURN, 1/2 W	BCD Radio Parts	1
RESISTOR: VARIABLE, 1/2 W, AUDIO	Fuji-Svea	12
RESISTOR: VARIABLE, 1/2 W, LINEAR	Fuji-Svea	67
RESISTOR: VARIABLE, 1/2 W, PC	Fuji-Svea	7
RESISTOR: VARIABLE, 1/2 WATT	Daytapro Electronics	12
RESISTOR: VARIABLE, 1/4 W, AUDIO	Mouser Electronics	7
RESISTOR: VARIABLE, 1/4 W, AUDIO	Fuji-Svea	7
RESISTOR: VARIABLE, 1/4 W, LINEAR	Circuit Specialists	7
RESISTOR: VARIABLE, 1/4 W, LINEAR	Fuji-Svea	21
RESISTOR: VARIABLE, 1/4 W, LINEAR	Mouser Electronics	40
RESISTOR: VARIABLE, 1/5 WATT, MINI	ETCO	5
RESISTOR: VARIABLE, 1/8 W, LINEAR	Mouser Electronics	7
RESISTOR: VARIABLE, 1/8 W, SUBMINI	Circuit Specialists	8
RESISTOR: VARIABLE, 1/8 W, SUBMINI	ETCO	7
RESISTOR: VARIABLE, 2 W, LINEAR	Jameco Electronics	7
RESISTOR: VARIABLE, 2 W, SCR ADJ	Star-Tronics	1
RESISTOR: VARIABLE, 25-300 WATT	Fair Radio Sales	70
RESISTOR: VARIABLE, 5W, SUBMINI	Circuit Specialists	18
RESISTOR: VARIABLE, 6mm DIA CERMET	Digi-Key	60
RESISTOR: VARIABLE, DUAL	Key Electronics	2
RESISTOR: VARIABLE, DUAL	ETCO	1
RESISTOR: VARIABLE, DUAL, AUDIO	Star-Tronics	1
RESISTOR: VARIABLE, DUAL, LINEAR	Mouser Electronics	9
RESISTOR: VARIABLE, DUAL, LINEAR	Circuit Specialists	3
RESISTOR: VARIABLE, DUAL, MINIATURE	Alpha Electronic Labs	1
RESISTOR: VARIABLE, FADER, 16 OHM	ETCO	2
RESISTOR: VARIABLE, FADER, 16 OHM	Semiconductors Surplus	1
RESISTOR: VARIABLE, FADER, 8 OHM	ETCO	5
RESISTOR: VARIABLE, FADER, L PAD	Fuji-Svea	2
RESISTOR: VARIABLE, FADER, L PAD	ETCO	3
RESISTOR: VARIABLE, FADER, L PAD	ETCO	9
RESISTOR: VARIABLE, FADER, T PAD	ETCO	2
RESISTOR: VARIABLE, FILM, MULTITURN	Mouser Electronics	61
RESISTOR: VARIABLE, LINEAR	BCD Radio Parts	20
RESISTOR: VARIABLE, LINEAR	Digital Research	2
RESISTOR: VARIABLE, LINEAR, 1 M OHM	Babylon Electronics	1
RESISTOR: VARIABLE, LINEAR, 1K OHM	Key Electronics	1
RESISTOR: VARIABLE, LINEAR, PC MT	Mouser Electronics	9
RESISTOR: VARIABLE, LINEAR, SUBMINI	Star-Tronics	1
RESISTOR: VARIABLE, MOTOR DRIVEN	ETCO	2
RESISTOR: VARIABLE, PC, SCR ADJ	Active Electronics	3
RESISTOR: VARIABLE, PC, SCR ADJ	ETCO	11

COMPONENT	COMPANY NAME	VARIETY STOCKED
RESISTOR: VARIABLE, PC, SCR ADJ	Digi-Key	26
RESISTOR: VARIABLE, PC, SCR ADJ	Sintec	6
RESISTOR: VARIABLE, PC, SCR ADJ	Mouser Electronics	94
RESISTOR: VARIABLE, PC, SCR ADJ	Jameco Electronics	24
RESISTOR: VARIABLE, PC, SCR ADJ	Star-Tronics	1
RESISTOR: VARIABLE, PC, SCR ADJ	Star-Tronics	2
RESISTOR: VARIABLE, PC, SCR ADJ	Circuit Specialists	8
RESISTOR: VARIABLE, PC, SCR ADJ	Mouser Electronics	1
RESISTOR: VARIABLE, PC, SCR ADJ	Fuji-Svea	56
RESISTOR: VARIABLE, PRECISN LINEAR	Star-Tronics	1
RESISTOR: VARIABLE, REVERSE AUDIO	Star-Tronics	1
RESISTOR: VARIABLE, SLIDE	ETCO	3
RESISTOR: VARIABLE, SLIDE, 10 K OHM	BCD Radio Parts	1
RESISTOR: VARIABLE, SLIDE, AUDIO	Circuit Specialists	5
RESISTOR: VARIABLE, SLIDE, AUDIO	Mouser Electronics	5
RESISTOR: VARIABLE, SLIDE, AUDIO	Fuji-Svea	6
RESISTOR: VARIABLE, SLIDE, DUAL	ETCO	1
RESISTOR: VARIABLE, SLIDE, DUAL	Semiconductors Surplus	1
RESISTOR: VARIABLE, SLIDE, LINEAR	Mouser Electronics	27
RESISTOR: VARIABLE, SLIDE, LINEAR	Circuit Specialists	8
RESISTOR: VARIABLE, SLIDE, LINEAR	Fuji-Svea	21
RESISTOR: VARIABLE, SLIDE, LINEAR	Semiconductors Surplus	4
RESISTOR: VARIABLE, SPEAKER, 8 OHMS	Mouser Electronics	2
RESISTOR: VARIABLE, STANDARD	ETCO	13
RESISTOR: VARIABLE, STANDARD	Star-Tronics	3
RESISTOR: VARIABLE, STANDARD	Key Electronics	2
RESISTOR: VARIABLE, STANDARD	Surplus Electronics	2
RESISTOR: VARIABLE, STANDARD RADIO	Circuit Specialists	13
RESISTOR: VARIABLE, STANDARD, AUDIO	Star-Tronics	1
RESISTOR: VARIABLE, STANDARD, AUDIO	Circuit Specialists	3
RESISTOR: VARIABLE, STANDARD, AUDIO	Mouser Electronics	19
RESISTOR: VARIABLE, STANDARD,LINEAR	Circuit Specialists	20
RESISTOR: VARIABLE, STANDARD,LINEAR	Mouser Electronics	84
RESISTOR: VARIABLE, THUMB-WHEEL	Key Electronics	1
RESISTOR: VARIABLE, THUMB-WHEEL	ETCO	14
RESISTOR: VARIABLE, THUMB-WHEEL	Star-Tronics	2
RESISTOR: VARIABLE, THUMB-WHEEL	Jameco Electronics	11
RESISTOR: VARIABLE, THUMB-WHEEL, PC	Digi-Key	26
RESISTOR: VARIABLE, THUMB-WHEEL, PC	Semiconductors Surplus	22
RESISTOR: VARIABLE, THUMB-WHEEL, PC	Circuit Specialists	12
RESISTOR: VARIABLE, THUMB-WHEEL, PC	Star-Tronics	1
RESISTOR: VARIABLE, THUMB-WHEEL, PC	Mouser Electronics	27
RESISTOR: VARIABLE, THUMB-WHEEL, PC	Sintec	6
RESISTOR: VARIABLE, THUMBWHL, AUDIO	Fuji-Svea	5
RESISTOR: VARIABLE, THUMBWHL,LINEAR	Fuji-Svea	5
RESISTOR: VARIABLE, TRANSISTR RADIO	Mouser Electronics	24
RESISTOR: VARIABLE, TRANSISTR RADIO	Circuit Specialists	2
RESISTOR: VARIABLE, TRIM, 10 OHM	Digital Research	1
RESISTOR: VARIABLE, TRIMMER	BCD Radio Parts	8
RESISTOR: VARIABLE, TRIMMER	Surplus Electronics	4
RESISTOR: VARIABLE, TRIMMER	Digital Research	6
RESISTOR: VARIABLE, TRIMMER, CERMET	Fair Radio Sales	31
RESISTOR: VARIABLE, TRIMMER, CERMET	BCD Radio Parts	9
RESISTOR: VARIABLE, WW, 25 WATT	Atlantic Surplus Sales	1
RESISTOR: VARIABLE, WW, 50K OHM	Aldelco	1
RESISTOR: VARIABLE, WW, 5K OHM	Key Electronics	1
RESISTOR: VARIABLE, WW, LINEAR, 1 W	Star-Tronics	1
RESISTOR: VARIABLE, WW, SCR ADJ	Star-Tronics	1
RESISTOR: VARIABLE, WW, STAND, LIN	Star-Tronics	1

COMPONENT	COMPANY NAME	VARIETY STOCKED
RESISTOR: WIREWOUND, 167 OHM, 1%	Star-Tronics	1
RESISTOR: WIREWOUND, 350 OHM, 5 W	Star-Tronics	1
RESISTOR: WIREWOUND, 5 OHM, 25 WAT	Star-Tronics	1
RESISTOR: WIREWOUND, 10%, 1 WATT	Proverty Electronics	
RESISTOR: WIREWOUND, 10%, 10 WATT	Mouser Electronics	38
RESISTOR: WIREWOUND, 10%, 3 WATT	Mouser Electronics	13
RESISTOR: WIREWOUND, 10%, 5 WATT	Mouser Electronics	40
RESISTOR: WIREWOUND, 5%, 10 WATT	Fuji-Svea	37
RESISTOR: WIREWOUND, 5%, 10 WATT	Mouser Electronics	15
RESISTOR: WIREWOUND, 5%, 5 WATT	Fuji-Svea	37
RESISTOR: WIREWOUND, 5%, 5 WATT	Mouser Electronics	20
RESISTOR: WIREWOUND, 50 WATT	Star-Tronics	4
RESISTOR: WIREWOUND, POWER	ETCO	4
RESONATOR: AUDIO, TELEPHONE TONE	ETCO	1
RESONATOR: CAVITY, 8.9-9.5 GHZ	Lectronic	1
RESONATOR: CERAMIC, 352.3 KHZ	BCD Radio Parts	1
RING HYBRID: COAXIAL, 1-2 GHZ	Lectronic	
ROD END: BALL SWIVEL	Small Parts	20
ROD: ROUND, DELRIN, 1/4-2" OD	Small Parts	6
ROD: ROUND, NYLON, ANTI-FRICTION	Small Parts	6
ROD: ROUND, PHENOLIC FIBRE, 1/16-1"	Small Parts	9
ROD: ROUND, TEFLON, 1/4-2" OD	Small Parts	6
ROD: THREADED, BRASS, NICKEL PLATED	Small Parts	5
ROD: THREADED, NYLON	Small Parts	8
ROD: THREADED, SS	Small Parts	12
ROD: THREADED, SS, METRIC SIZE	Small Parts	6
ROD: THREADED, TITANIUM	Small Parts	1
ROPE: ANTENNA GUY, NYLON	G & C Communications	1
ROPE: ANTENNA GUY, NYLON	Van Gorden Engineering	1
SCR:	See Transistor: SCR	
SCREEN: NYLON, FINE MESH	Small Parts	19
SCREEN: POLYPROPYLENE, FINE MESH	Small Parts	6
SCREEN: WIRE, SS, FINE MESH	Small Parts	19
SEAL: HYDRAULIC, O-RING, RUBBER	Small Parts	31
SELSYN: SYNCHRO CONTROL	Fair Radio Sales	16
SERVO: ASSEMBLY, 115 VAC	Star-Tronics	1
SHAFT COLLAR: METAL, 1/8-1" BORE	Small Parts	29
SHAFT:	See Also Rod:	
SHAFT: FLEXIBLE, 1/8 & 1/4" END	Small Parts	14
SHAFT: STEEL, 1/16 TO 2" DIAMETER	Small Parts	22
SHAFT: STEEL, 1/4" DIAMETER	Star-Tronics	1
SHEET METAL: ALUMINUM, .032-.375" T	Small Parts	6
SHEET METAL: ALUMINUM, 1/8x12x18"	Star-Tronics	1
SHEET METAL: BRASS, .020-.250" T	Small Parts	6
SHEET METAL: COPPER, .021-.1875" T	Small Parts	5
SHEET METAL: PERFORATED, SS	Small Parts	4

COMPONENT	COMPANY NAME	VARIETY STOCKED
SHIM STOCK: BRASS, .001-.031"	Small Parts	15
SHIM STOCK: STEEL, .001-.031"	Small Parts	15
SIDAC:	See Transistor: SIDAC	
SOCKET TERMINAL TURRET: TUBE, 7 PIN	ETCO	1
SOCKET: AC ELECTRICAL	See Connector: AC Electrical	
SOCKET: CATHODE RAY TUBE	Fair Radio Sales	3
SOCKET: CATHODE RAY TUBE	ETCO	5
SOCKET: CONTACT PINS, BERG	BCD Radio Parts	1
SOCKET: CONTACT PINS, BERG MINISERT	Sintec	1
SOCKET: CRYSTAL	C-W Crystals	4
SOCKET: CRYSTAL	International Crystal	5
SOCKET: CRYSTAL	Surplus Electronics	3
SOCKET: CRYSTAL, FT 243	ETCO	1
SOCKET: CRYSTAL, HC 25/U	Spectrum International	2
SOCKET: CRYSTAL, HC 25/U	Aldelco	2
SOCKET: CRYSTAL, HC 25/U, MULTIPLE	Mouser Electronics	6
SOCKET: CRYSTAL, HC 6/U	Spectrum International	1
SOCKET: CRYSTAL, HC 6/U	ETCO	1
SOCKET: CRYSTAL, STEATITE, MILLEN	Radiokit	4
SOCKET: CRYSTAL, STEATITE, TYPE A&B	Fuji-Svea	2
SOCKET: CRYSTAL, XM-107-S04	Spectrum International	1
SOCKET: DISPLAY, HORIZONTAL	Sintec	29
SOCKET: DISPLAY, HORIZONTAL	Digi-Key	2
SOCKET: DISPLAY, VERTISOCKET	Digi-Key	5
SOCKET: DISPLAY, VERTISOCKET	Mouser Electronics	11
SOCKET: DISPLAY, VERTISOCKET	Circuit Specialists	2
SOCKET: DISPLAY, VERTISOCKET	Sintec	10
SOCKET: IC, ELEVATOR, .25" TO 1.25"	Sintec	11
SOCKET: IC, ELEVATOR, .55" TO 1.04"	Mouser Electronics	4
SOCKET: IC, LOW PROFILE	Active Electronics	18
SOCKET: IC, MOLEX PIN NEST	Jameco Electronics	4
SOCKET: IC, MOLEX PIN NEST	Quest Electronics	3
SOCKET: IC, MOLEX PINS	Quest Electronics	1
SOCKET: IC, MOLEX PINS	Jameco Electronics	1
SOCKET: IC, MOLEX PINS	Digi-Key	1
SOCKET: IC, SINGLE PIN, WIRE WRAP	Mouser Electronics	6
SOCKET: IC, SOCKET STRIP, SAMTEC	Mouser Electronics	2
SOCKET: IC, SOCKET STRIP, SOLDER	Sintec	21
SOCKET: IC, SOCKET STRIP, WIRE WRAP	Sintec	21
SOCKET: IC, SOLDER TAIL	Surplus Electronics	6
SOCKET: IC, SOLDER TAIL	Babylon Electronics	6
SOCKET: IC, SOLDER TAIL	Semiconductors Surplus	9
SOCKET: IC, SOLDER, HI-REL	Mouser Electronics	18
SOCKET: IC, SOLDER, LOW PROFILE	Sintec	9
SOCKET: IC, SOLDER, LOW PROFILE	Quest Electronics	10
SOCKET: IC, SOLDER, LOW PROFILE	ETCO	4
SOCKET: IC, SOLDER, LOW PROFILE	Aldelco	7
SOCKET: IC, SOLDER, LOW PROFILE	Jameco Electronics	10
SOCKET: IC, SOLDER, LOW PROFILE	Digi-Key	18
SOCKET: IC, SOLDER, LOW PROFILE	Fuji-Svea	9
SOCKET: IC, SOLDER, LOW PROFILE	Mouser Electronics	21
SOCKET: IC, SOLDER, LOW PROFILE	Daytapro Electronics	5
SOCKET: IC, SOLDER, LOW PROFILE	Circuit Specialists	8
SOCKET: IC, SOLDER, LOW PROFILE	Fair Radio Sales	7
SOCKET: IC, SOLDER, LOW PROFILE	B.G. Micro	8

COMPONENT	COMPANY NAME	VARIETY STOCKED
SOCKET: IC, SOLDER, SCREW MACH PINS	Circuit Specialists	5
SOCKET: IC, SOLDER, SCREW MACH PINS	Digi-Key	9
SOCKET: IC, SOLDER, SCREW MACH PINS	Sintec	10
SOCKET: IC, SOLDER, SCREW MACH PINS	Active Electronics	9
SOCKET: IC, SOLDER, SCREW MACH PINS	Star-Tronics	1
SOCKET: IC, SOLDER, STANDARD	Fair Radio Sales	4
SOCKET: IC, SOLDER, STANDARD	ETCO	2
SOCKET: IC, SOLDER, STANDARD	Digital Research	7
SOCKET: IC, SOLDER, STANDARD	Quest Electronics	6
SOCKET: IC, SOLDER, STANDARD	Jameco Electronics	15
SOCKET: IC, SOLDER, STANDARD	Electronic Marketplace	9
SOCKET: IC, SOLDER, STANDARD	BCD Radio Parts	10
SOCKET: IC, SOLDER, STANDARD	Key Electronics	4
SOCKET: IC, SOLDER, STANDARD	Star-Tronics	2
SOCKET: IC, TO-5, 3 TO 10 PIN	Jameco Electronics	4
SOCKET: IC, TO-5, 3 TO 12 PIN	Sintec	6
SOCKET: IC, TO-5, 8-PIN	Mouser Electronics	1
SOCKET: IC, WIRE WRAP	Fuji-Svea	5
SOCKET: IC, WIRE WRAP	Surplus Electronics	2
SOCKET: IC, WIRE WRAP	Jameco Electronics	13
SOCKET: IC, WIRE WRAP	Circuit Specialists	5
SOCKET: IC, WIRE WRAP	Sintec	18
SOCKET: IC, WIRE WRAP	Mouser Electronics	18
SOCKET: IC, WIRE WRAP	Electronic Marketplace	9
SOCKET: IC, WIRE WRAP	Fair Radio Sales	1
SOCKET: IC, WIRE WRAP	Digi-Key	18
SOCKET: IC, WIRE WRAP	Quest Electronics	9
SOCKET: IC, WIRE WRAP	Babylon Electronics	5
SOCKET: IC, WIRE WRAP	BCD Radio Parts	2
SOCKET: IC, WIRE WRAP	Active Electronics	9
SOCKET: IC, WIRE WRAP	ETCO	1
SOCKET: IC, WIRE WRAP	Semiconductors Surplus	3
SOCKET: IC, WIRE WRAP, GOLD PLATE	Digital Research	2
SOCKET: IC, WIRE WRAP, SCR MACH PIN	Circuit Specialists	5
SOCKET: IC, WIRE WRAP, SCR MACH PIN	Sintec	10
SOCKET: IC, WIRE WRAP, SCR MACH PIN	Digi-Key	9
SOCKET: IC, ZERO FORCE INSERTION	Quest Electronics	8
SOCKET: IC, ZERO FORCE INSERTION	Jameco Electronics	16
SOCKET: IC, ZERO FORCE INSERTION	Mouser Electronics	6
SOCKET: IC, ZERO FORCE INSERTION	Sintec	8
SOCKET: INCANDESCENT LAMP	Mouser Electronics	10
SOCKET: INCANDESCENT LAMP	ETCO	10
SOCKET: INCANDESCENT LAMP, AUTO	ETCO	1
SOCKET: INCANDESCENT LAMP, BAYONET	Surplus Electronics	2
SOCKET: INCANDESCENT LAMP, BI-PIN	ETCO	1
SOCKET: INCANDESCENT LAMP, SCREW B	Surplus Electronics	1
SOCKET: INCANDESCENT LAMP,T-3 1/4,5	Surplus Electronics	1
SOCKET: PIN-LINE, SOLDER & WRAP	Digi-Key	2
SOCKET: PIN-LINE, SOLDER & WRAP	Circuit Specialists	2
SOCKET: RELAY	Mouser Electronics	4
SOCKET: RELAY	Surplus Electronics	4
SOCKET: RELAY, 12 PIN	Star-Tronics	1
SOCKET: RELAY, 9 PIN	Star-Tronics	2
SOCKET: RELAY, CIRCULAR PIN CONFIG	Digi-Key	6
SOCKET: RELAY, RECTANGULAR PIN CONF	Digi-Key	3
SOCKET: TRANSISTOR	Key Electronics	1
SOCKET: TRANSISTOR	ETCO	6
SOCKET: TRANSISTOR, PC MT	Mouser Electronics	7
SOCKET: TRANSISTOR, POWER, TO-3	Jameco Electronics	1

COMPONENT	COMPANY NAME	VARIETY STOCKED
SOCKET: TRANSISTOR, POWER, TO-3	Digital Research	1
SOCKET: TRANSISTOR, POWER, TO-3	Semiconductors Surplus	1
SOCKET: TRANSISTOR, POWER, TO-3,-66	Mouser Electronics	4
SOCKET: TRANSISTOR, POWER, TO-3,INS	Surplus Electronics	1
SOCKET: TRANSISTOR, TO-3	BCD Radio Parts	1
SOCKET: TRANSISTOR, TO-5	Star-Tronics	1
SOCKET: TRANSISTOR, TO-5	Jameco Electronics	1
SOCKET: TRANSISTOR, TO-66, INS MT	Surplus Electronics	1
SOCKET: TRANSISTOR, TO-66, INS MT	Star-Tronics	1
SOCKET: TRANSISTOR, UNIVERSAL, LP	Mouser Electronics	3
SOCKET: TRANSISTOR, WIRE WRAP	Mouser Electronics	2
SOCKET: TUBE, 3-1000	Amp Supply	1
SOCKET: TUBE, 4-1000	Amp Supply	1
SOCKET: TUBE, 4-125A,4-400A,JOHNSON	Fair Radio Sales	1
SOCKET: TUBE, 572	Amp Supply	1
SOCKET: TUBE, 811	Amp Supply	1
SOCKET: TUBE, 8877	Amp Supply	1
SOCKET: TUBE, BAKELITE, 7&9 PIN	Fuji-Svea	2
SOCKET: TUBE, BAKELITE, 7-PIN	Mouser Electronics	1
SOCKET: TUBE, CERAMIC, 4,5,7,9 PIN	Radiokit	4
SOCKET: TUBE, CERAMIC, 7&9 PIN MINI	Fair Radio Sales	2
SOCKET: TUBE, CERAMIC, JOHNSON XMIT	Fair Radio Sales	11
SOCKET: TUBE, CERAMIC, OCTAL	Fair Radio Sales	1
SOCKET: TUBE, CERAMIC, TYPE 4,5,6,7	Fair Radio Sales	4
SOCKET: TUBE, GRID BYPASS, 4X150-	Fair Radio Sales	1
SOCKET: TUBE, HIGH VOLTAGE, OCTAL	Surplus Electronics	1
SOCKET: TUBE, HIGH VOLTAGE, OCTAL	ETCO	1
SOCKET: TUBE, LOCTAL, MOLDED	Fair Radio Sales	1
SOCKET: TUBE, MOULDED, 7&9 PIN	ETCO	2
SOCKET: TUBE, MOULDED, 7&9 PIN MINI	Fair Radio Sales	2
SOCKET: TUBE, MOULDED, 7,9,12 PIN	Star-Tronics	5
SOCKET: TUBE, MOULDED, OCTAL	Fair Radio Sales	2
SOCKET: TUBE, MOULDED, TYPE 4,5,6,7	Fair Radio Sales	4
SOCKET: TUBE, PLASTIC, OCTAL	ETCO	1
SOCKET: TUBE, PRINTED CKT, 9 PIN	ETCO	1
SOCKET: TUBE, SECONDARY	ETCO	5
SOCKET: TUBE, STEATITE, 7&9 PIN	Fuji-Svea	2
SOCKET: TUBE, STEATITE, US TYPE	Fuji-Svea	1
SOCKET: TUBE, STEATITE, UX TYPE	Fuji-Svea	1
SOCKET: TUBE, STEATITE, UY TYPE	Fuji-Svea	1
SOCKET: TUBE, STEATITE, UZ TYPE	Fuji-Svea	1
SOCKET: TUBE, TEST, 7 PIN	ETCO	2
SOCKET: TUBE, TRANSMITTING, EIMAC	MHZ Electronics	15
SOCKET: TUBE, TRANSMITTING, JOHNSON	MHZ Electronics	3
SOCKET: TUBE, TRANSMITTING, JOHNSON	Amp Supply	1
SOCKET: TUBE, TRANSMITTING, JOHNSON	Semiconductors Surplus	5
SOCKET: TUBE, WAFER, 7&9 PIN	ETCO	2
SOCKET: VIDICON TUBE	ATV Research	1
SOLAR CELL:	Edmund Scientific	11
SOLAR CELL:	Jameco Electronics	1
SOLAR CELL: PANEL	Edmund Scientific	7
SOLAR CELL: PANEL, 3,6,9 V, 50 MA	ETCO	1
SOLAR CELL: PANEL, 5 OR 10 V	Jameco Electronics	1
SOLAR CELL: PANEL, 6 V, 100 MA	ETCO	1
SOLAR CELL: PANEL, PHOTOVOLTIAC	Schnobel	2
SOLAR CELL: SILICON	Mouser Electronics	2
SOLDER FLUX: ALUMINUM TO ALUMINUM	Small Parts	1

COMPONENT	COMPANY NAME	VARIETY STOCKED
SOLDER FLUX: COMMON	Small Parts	1
SOLDER FLUX: LIQUID	Mouser Electronics	1
SOLDER FLUX: PASTE	ETCO	1
SOLDER FLUX: RESIN, KESTER	Mouser Electronics	1
SOLDER FLUX: ROSIN	Fuji-Svea	1
SOLDER FLUX: SILVER BRAZING	Small Parts	2
SOLDER WICK:	Daytapro Electronics	1
SOLDER WICK:	Mouser Electronics	4
SOLDER WICK:	Jameco Electronics	3
SOLDER WICK:	Digi-Key	10
SOLDER WICK:	Sintec	2
SOLDER: ALUMINUM, ALSOLDER 500	Small Parts	1
SOLDER: DISPENSER PACK, ROSIN CORE	Jameco Electronics	1
SOLDER: DISPENSER PACK, ROSIN CORE	Fuji-Svea	1
SOLDER: MULTICORE	Digi-Key	4
SOLDER: POCKET-PACK	Daytapro Electronics	1
SOLDER: RESIN CORE, 60/40, KESTER	BCD Radio Parts	1
SOLDER: RESIN CORE, KESTER	Fuji-Svea	1
SOLDER: RESIN CORE, KESTER	Mouser Electronics	7
SOLDER: RESIN CORE, MULTICORE	Mouser Electronics	5
SOLDER: ROSIN CORE	Fair Radio Sales	1
SOLDER: ROSIN CORE	Small Parts	4
SOLDER: ROSIN CORE	Jameco Electronics	1
SOLDER: ROSIN CORE, AE 16	Sintec	6
SOLDER: SILVER BRAZING, SAFETY-SILV	Small Parts	10
SOLENOID: 110 VAC, 1 1/8" THROW	ETCO	1
SOLENOID: 12 VDC, 1/4" THROW	ETCO	1
SOLENOID: 24 V, 1/8" THROW	Surplus Electronics	1
SOLENOID: 24 VDC, 1/2" THROW	BCD Radio Parts	1
SOLENOID: 34 VDC	ETCO	1
SPEAKER GRILL:	ETCO	6
SPEAKER GRILL: 3X3"	Aldelco	1
SPEAKER:	Semiconductors Surplus	4
SPEAKER: AUDIO	Aldelco	2
SPEAKER: AUDIO, 200 OHM	BCD Radio Parts	1
SPEAKER: AUDIO, 200 OHM	Digital Research	1
SPEAKER: AUDIO, 3.2 OHM, 3.25" SQ	Surplus Electronics	1
SPEAKER: AUDIO, 32 OHM	ETCO	1
SPEAKER: AUDIO, 45 OHM, 5" DIAMETER	Star-Tronics	1
SPEAKER: AUDIO, 800 OHM VOICE COIL	ETCO	1
SPEAKER: AUDIO, CERAMIC, MINIATURE	Mouser Electronics	10
SPEAKER: AUDIO, HI-FI	ETCO	7
SPEAKER: AUDIO, HI-FI, TWIN CONE	Mouser Electronics	1
SPEAKER: AUDIO, INTERCOM, 45 OHM	ETCO	7
SPEAKER: AUDIO, LOW PROFILE,DYNAMIC	Mouser Electronics	3
SPEAKER: AUDIO, MINIATURE	Jameco Electronics	1
SPEAKER: AUDIO, MINIATURE	Surplus Electronics	1
SPEAKER: AUDIO, MINIATURE	BCD Radio Parts	1
SPEAKER: AUDIO, MINIATURE	ETCO	22
SPEAKER: AUDIO, MINIATURE	Quest Electronics	2
SPEAKER: AUDIO, MINIATURE, ALNICO	Mouser Electronics	22
SPEAKER: AUDIO, MINIATURE, SQUARE	Jameco Electronics	1
SPEAKER: AUDIO, MYLAR CONE	ETCO	3
SPEAKER: AUDIO, MYLAR CONE, MOUSER	Mouser Electronics	2

COMPONENT	COMPANY NAME	VARIETY STOCKED
SPEAKER: AUDIO, STANDARD REPLACEMNT	ETCO	2
SPEAKER: EARPHONE	ETCO	4
SPEAKER: EARPHONE	BCD Radio Parts	1
SPEAKER: ULTRASONIC TRANSDUCER	Semiconductors Surplus	1
SPEAKER: ULTRASONIC TRANSDUCER	Surplus Electronics	1
SPEAKER: ULTRASONIC TRANSDUCER	Circuit Specialists	1
SPEAKER: ULTRASONIC TRANSDUCER	Mouser Electronics	1
SPLITTER: RF, 16 WAY, .5-125 MHZ	Mini-Circuits Laboratory	1
SPLITTER: RF, 2 WAY, 420-450 MHZ	RIW Products	1
SPLITTER: RF, 2 WAY, DC TO 2 GHZ	Elcom Systems	1
SPLITTER: RF, 2 WAY, TO 4.2 GHZ	Mini-Circuits Laboratory	46
SPLITTER: RF, 24 WAY, .2-100 MHZ	Mini-Circuits Laboratory	1
SPLITTER: RF, 3 WAY, DC TO 500 MHZ	Elcom Systems	1
SPLITTER: RF, 3 WAY, TO 750 MHZ	Mini-Circuits Laboratory	14
SPLITTER: RF, 4 WAY, 420-450 MHZ	RIW Products	1
SPLITTER: RF, 4 WAY, TO 1 GHZ	Mini-Circuits Laboratory	15
SPLITTER: RF, 6 WAY, TO 175 MHZ	Mini-Circuits Laboratory	2
SPLITTER: RF, 8 WAY, 420-450 MHZ	RIW Products	1
SPLITTER: RF, 8 WAY, DC TO 500 MHZ	Elcom Systems	1
SPLITTER: RF, 8 WAY, TO 700 MHZ	Mini-Circuits Laboratory	9
SPLITTER: RF, MICROWAVE	Lectronic	
SPLITTER: RF, VHF, UHF, TV	ETCO	6
SPRING: COMPRESSION, BERYLLIUM COPR	Small Parts	48
SPRING: FLAT, STEEL	Small Parts	8
STRIPS: BRASS, .016-.032" THICK	Small Parts	12
STUB TUNER: COAXIAL, MICROWAVE	Lectronic	
STUB TUNER: WAVEGUIDE, MICROWAVE	Lectronic	
SUCTION CUP: MOUNTING STUD	ETCO	1
SWITCH CAPS: COLORED, BAT HANDLE	Star-Tronics	ASST
SWITCH: 15 BUTTON, MAKE 1, BREAK 1	Fair Radio Sales	1
SWITCH: 3 BUTTON, INTERLOCKING	Star-Tronics	1
SWITCH: 5 BUTTON, MAKE 1, BREAK 1	Star-Tronics	1
SWITCH: AIR PRESSURE, DIFFERENTIAL	Edmund Scientific	1
SWITCH: AIRCRAFT TYPE	Fair Radio Sales	4
SWITCH: BATON SLIDE, DPDT	Digital Research	1
SWITCH: DIP	Active Electronics	10
SWITCH: DIP	BCD Radio Parts	5
SWITCH: DIP	Alpha Electronic Labs	
SWITCH: DIP	Mouser Electronics	9
SWITCH: DIP	Quest Electronics	7
SWITCH: DIP, MULTI-POSITION	B.G. Micro	4
SWITCH: DIP, ROCKER	Fair Radio Sales	1
SWITCH: DIP, ROCKER	Jameco Electronics	10
SWITCH: DIP, SLIDE	Jameco Electronics	3
SWITCH: DIP, SLIDE	Surplus Electronics	4
SWITCH: DIP, SLIDE	Sintec	8
SWITCH: DIP, SLIDE	Babylon Electronics	1
SWITCH: DIP, SLIDE	ETCO	6
SWITCH: FLOOR AREA MAT	ETCO	1
SWITCH: FLOW SENSING, AIR	ETCO	1
SWITCH: FOOT PEDAL	ETCO	3
SWITCH: FOOT PEDAL, 20 AMP	Fair Radio Sales	1

COMPONENT	COMPANY NAME	VARIETY STOCKED
SWITCH: JUMPER CABLE INTRA-SWITCH	Jameco Electronics	2
SWITCH: KNIFE, DPDT	Fuji-Svea	1
SWITCH: KNIFE, PORCELAIN BASE	ETCO	4
SWITCH: LEAF, MINIATURE	ETCO	2
SWITCH: LEAF, SPDT	Surplus Electronics	1
SWITCH: LEVER	Fair Radio Sales	8
SWITCH: LEVER, TELEPHONE	ETCO	1
SWITCH: MERCURY TILT	ETCO	2
SWITCH: MICRO	Fair Radio Sales	15
SWITCH: MICRO	ETCO	5
SWITCH: MICRO	Star-Tronics	1
SWITCH: MICRO	Surplus Electronics	1
SWITCH: PADDLE, COMPUTER	BCD Radio Parts	1
SWITCH: PADDLE, IMSAI COMPUTER	Digital Research	1
SWITCH: PLUNGER	BCD Radio Parts	1
SWITCH: PLUNGER ACTIVATED	ETCO	2
SWITCH: PLUNGER ACTUATED	Digital Research	1
SWITCH: PRESSURE, OMNI-DIRECTIONAL	ETCO	2
SWITCH: PUSHBUTTON	Surplus Electronics	2
SWITCH: PUSHBUTTON	Circuit Specialists	7
SWITCH: PUSHBUTTON	ETCO	27
SWITCH: PUSHBUTTON	Fair Radio Sales	9
SWITCH: PUSHBUTTON BANK	Digital Research	1
SWITCH: PUSHBUTTON BANK, CANCELLING	BCD Radio Parts	4
SWITCH: PUSHBUTTON, 3-GANG	Aldelco	1
SWITCH: PUSHBUTTON, COLOR INDICATOR	Star-Tronics	1
SWITCH: PUSHBUTTON, ILLUMINATED	Aldelco	1
SWITCH: PUSHBUTTON, MOMENTARY	Star-Tronics	1
SWITCH: PUSHBUTTON, MOMENTARY	Active Electronics	2
SWITCH: PUSHBUTTON, MOMENTARY	Quest Electronics	3
SWITCH: PUSHBUTTON, MOMENTARY	Digi-Key	2
SWITCH: PUSHBUTTON, MOMENTARY	Jameco Electronics	5
SWITCH: PUSHBUTTON, MOMENTARY	Mouser Electronics	45
SWITCH: PUSHBUTTON, MOMENTARY, NC	BCD Radio Parts	1
SWITCH: PUSHBUTTON, NC	Digital Research	1
SWITCH: PUSHBUTTON, ON/OFF LATCH	Quest Electronics	2
SWITCH: PUSHBUTTON, ON/OFF LATCH	Jameco Electronics	2
SWITCH: PUSHBUTTON, ON/OFF LATCH	Digi-Key	1
SWITCH: PUSHBUTTON, ON/OFF LATCH	Mouser Electronics	18
SWITCH: PUSHBUTTON, SNAP ACTING	Circuit Specialists	2
SWITCH: PUSHBUTTON, STRIP MT	Fair Radio Sales	2
SWITCH: PUSHBUTTON, SUBMINIATURE	Semiconductors Surplus	1
SWITCH: REED, MAGNET ACTIVATED	Mouser Electronics	3
SWITCH: REED, MAGNET ACTIVATED	Semiconductors Surplus	1
SWITCH: REED, MAGNET ACTIVATED	Star-Tronics	1
SWITCH: REED, MAGNET ACTIVATED	ETCO	4
SWITCH: RF	See Also Relay: Coax	
SWITCH: RF, ELECTRONIC, TO 2.5 GHZ	Mini-Circuits Laboratory	8
SWITCH: ROCKER	Surplus Electronics	5
SWITCH: ROCKER	ETCO	7
SWITCH: ROCKER	Digi-Key	2
SWITCH: ROCKER	Active Electronics	3
SWITCH: ROCKER	Star-Tronics	1
SWITCH: ROCKER	BCD Radio Parts	4
SWITCH: ROCKER	Mouser Electronics	15
SWITCH: ROCKER	Circuit Specialists	5
SWITCH: ROCKER	Alpha Electronic Labs	2
SWITCH: ROCKER	Fair Radio Sales	4
SWITCH: ROCKER SLIDE	Digital Research	2

COMPONENT	COMPANY NAME	VARIETY STOCKED
SWITCH: ROCKER, 3PDT, MINI	Digital Research	1
SWITCH: ROTARY	ETCO	22
SWITCH: ROTARY	BCD Radio Parts	5
SWITCH: ROTARY	Digital Research	3
SWITCH: ROTARY	Daytapro Electronics	6
SWITCH: ROTARY	Active Electronics	6
SWITCH: ROTARY	Surplus Electronics	6
SWITCH: ROTARY, 40 POSITION, CB	Aldelco	1
SWITCH: ROTARY, 40 POSITION, CB	BCD Radio Parts	1
SWITCH: ROTARY, 40 POSITION, CB	Digital Research	1
SWITCH: ROTARY, ANTENNA, 100 WATT	Fair Radio Sales	1
SWITCH: ROTARY, CERAMIC	Star-Tronics	1
SWITCH: ROTARY, CONTINUOUS TURN	Fair Radio Sales	1
SWITCH: ROTARY, MULTI-CONT,GOLDWELL	Fuji-Svea	33
SWITCH: ROTARY, MULTI-POS, HI-REL	Mouser Electronics	12
SWITCH: ROTARY, MULTI-POSITION	Jameco Electronics	2
SWITCH: ROTARY, MULTI-POSITION	Circuit Specialists	9
SWITCH: ROTARY, MULTI-POSITION	Fair Radio Sales	14
SWITCH: ROTARY, MULTI-POSITION	Mouser Electronics	77
SWITCH: ROTARY, MULTI-POSITION, PC	Mouser Electronics	36
SWITCH: ROTARY, RF, HP	Amp Supply	3
SWITCH: ROTARY, RF, HP, MILLEN	Radiokit	2
SWITCH: ROTARY, STEPPING, 50 VDC	ETCO	1
SWITCH: SECURITY, KEY	Mouser Electronics	1
SWITCH: SECURITY, KEY	BCD Radio Parts	1
SWITCH: SECURITY, KEY	ETCO	2
SWITCH: SLIDE	Surplus Electronics	2
SWITCH: SLIDE	ETCO	17
SWITCH: SLIDE TOGGLE	Circuit Specialists	4
SWITCH: SLIDE, 3 POSISTION, 2P3T	Star-Tronics	2
SWITCH: SLIDE, 5 POSITION, SP5T	Star-Tronics	1
SWITCH: SLIDE, HEAVY DUTY	BCD Radio Parts	4
SWITCH: SLIDE, HEAVY DUTY	Digital Research	3
SWITCH: SLIDE, HORIZONTAL MOUNT, PC	Digi-Key	1
SWITCH: SLIDE, ILLUMINATED, 6.3 V	ETCO	1
SWITCH: SLIDE, MICROMINI	Digi-Key	2
SWITCH: SLIDE, MINIATURE	Quest Electronics	2
SWITCH: SLIDE, MINIATURE	Star-Tronics	1
SWITCH: SLIDE, MINIATURE	Key Electronics	1
SWITCH: SLIDE, MINIATURE	Mouser Electronics	14
SWITCH: SLIDE, MINIATURE	Digi-Key	5
SWITCH: SLIDE, MINIATURE, PC MT	Active Electronics	5
SWITCH: SLIDE, RF ISOLATION, LP	Mouser Electronics	1
SWITCH: SLIDE, SIDE ACTUATED	Digi-Key	3
SWITCH: SLIDE, SPRING RETURN	Circuit Specialists	3
SWITCH: SLIDE, SPRING RETURN	Digi-Key	1
SWITCH: SLIDE, STANDARD	Jameco Electronics	1
SWITCH: SLIDE, STANDARD	Mouser Electronics	15
SWITCH: SLIDE, STANDARD	Digi-Key	8
SWITCH: SLIDE, STANDARD, 3 POSITION	Digi-Key	2
SWITCH: SLIDE, STANDARD, 4 POSITION	Digi-Key	1
SWITCH: SLIDE, STANDARD, DETENT	Circuit Specialists	9
SWITCH: SLIDE, SUBMINIATURE	Mouser Electronics	5
SWITCH: SLIDE, SUBMINIATURE	Jameco Electronics	1
SWITCH: SLIDE, SUBMINIATURE, DETENT	Circuit Specialists	4
SWITCH: SLIDE, SUBMINIATURE, PC MT	Mouser Electronics	1
SWITCH: SLIDE, ULTRAMINIATURE	Alpha Electronic Labs	1
SWITCH: SLIDE, VERTICAL MOUNT, PC	Digi-Key	1
SWITCH: SNAP ACTION. MINIATURE	Mouser Electronics	5

COMPONENT	COMPANY NAME	VARIETY STOCKED
SWITCH: STEPPING, 24 V DC	Fair Radio Sales	1
SWITCH: TAP, MULTI-CONTACT	Fair Radio Sales	8
SWITCH: THERMAL, 110 DEG C ACTIVATE	Star-Tronics	1
SWITCH: THERMAL, 135 DEG C CLOSURE	ETCO	1
SWITCH: THERMAL, 135 DEG F CLOSURE	ETCO	1
SWITCH: THUMBWHEEL, BCD	Jameco Electronics	2
SWITCH: THUMBWHEEL, BCD	Semiconductors Surplus	2
SWITCH: THUMBWHEEL, BCD	Circuit Specialists	1
SWITCH: THUMBWHEEL, BCD	Semiconductors Surplus	1
SWITCH: THUMBWHEEL, DECIMAL	Jameco Electronics	2
SWITCH: THUMBWHEEL, HEXADECIMAL	Jameco Electronics	3
SWITCH: TOGGLE	Surplus Electronics	13
SWITCH: TOGGLE	Fair Radio Sales	21
SWITCH: TOGGLE	Circuit Specialists	8
SWITCH: TOGGLE	Fuji-Svea	2
SWITCH: TOGGLE	ETCO	22
SWITCH: TOGGLE, CENTER OFF	Key Electronics	1
SWITCH: TOGGLE, FLAT LEVER	Jameco Electronics	5
SWITCH: TOGGLE, ILLUMINATED, LED	Mouser Electronics	1
SWITCH: TOGGLE, ILLUMINATED, LED	ETCO	3
SWITCH: TOGGLE, MICRO-MINI	Digital Research	1
SWITCH: TOGGLE, MINIATURE	Digi-Key	1
SWITCH: TOGGLE, MINIATURE	Active Electronics	5
SWITCH: TOGGLE, MINIATURE	Semiconductors Surplus	2
SWITCH: TOGGLE, MINIATURE	BCD Radio Parts	10
SWITCH: TOGGLE, MINIATURE	Circuit Specialists	3
SWITCH: TOGGLE, MINIATURE	Fuji-Svea	4
SWITCH: TOGGLE, MINIATURE	Mouser Electronics	76
SWITCH: TOGGLE, PADDLE	Jameco Electronics	1
SWITCH: TOGGLE, PC MOUNT	Jameco Electronics	4
SWITCH: TOGGLE, RIGHT ANGLE PC MT	Active Electronics	2
SWITCH: TOGGLE, STANDARD DUTY	Circuit Specialists	6
SWITCH: TOGGLE, STANDARD SIZE	Mouser Electronics	25
SWITCH: TOGGLE, SUBMINIATURE	Quest Electronics	4
SWITCH: TOGGLE, SUBMINIATURE	Mouser Electronics	20
SWITCH: TOGGLE, SUBMINIATURE	Jameco Electronics	4
SWITCH: TOUCH CAPSULE, .005" MOTION	Sintec	1
SWITCH: TOUCH CAPSULE, .4mm TRAVEL	Digi-Key	3
SWITCH: WATER-TIGHT, TOGGLE	Surplus Electronics	2
SWITCH: WAVEGUIDE, 1.12-170 GHZ	Lectronic	
TAPE: ABRASIVE	Small Parts	12
TAPE: COPPER, RF SHEILDING, 3M	Fox Tango Corporation	2
TAPE: ELECTRICAL, PLASTIC, BLACK	Fuji-Svea	1
TAPE: ELECTRICAL, PLASTIC, BLACK	Mouser Electronics	1
TAPE: FOIL, RECORDING CUE, ALUMINUM	ETCO	1
TAPE: RECORDING TAPE SPLICE	ETCO	2
TEFLON: BALLS, 3/32 TO 1" DIAMETER	Small Parts	9
TEFLON: BAR, RECTANGULAR, SOLID	Frank Wirt Electronics & TV	4
TEFLON: ROD, ROUND, SOLID	Frank Wirt Electronics & TV	6
TEFLON: ROD, ROUND, SOLID	Small Parts	6
TEFLON: SHEET, BARE	Frank Wirt Electronics & TV	9
TELETYPE: PARTS	Atlantic Surplus Sales	
TELETYPE: PARTS	Typetronics	
TERMINAL BLOCK JUMPER CLIP: CINCH	ETCO	1

COMPONENT	COMPANY NAME	VARIETY STOCKED
TERMINAL BLOCK:	Daytapro Electronics	5
TERMINAL BLOCK: AIRCRAFT TYPE	Fair Radio Sales	5
TERMINAL BLOCK: BARRIER, MILLEN	Radiokit	1
TERMINAL BLOCK: BARRIER, MODULAR	Sintec	23
TERMINAL BLOCK: BARRIER, SCREW TYPE	ETCO	4
TERMINAL BLOCK: BARRIER, SCREW TYPE	Star-Tronics	1
TERMINAL BLOCK: BARRIER, SCREW TYPE	Active Electronics	18
TERMINAL BLOCK: BARRIER, SCREW TYPE	Semiconductors Surplus	5
TERMINAL BLOCK: BARRIER, TRW	Mouser Electronics	12
TERMINAL BLOCK: COMPRESSION, 2 COND	Mouser Electronics	4
TERMINAL BLOCK: SCREW TYPE	Semiconductors Surplus	3
TERMINAL BLOCK: SUBMINIATURE	Digital Research	3
TERMINAL BLOCK: SUBMINIATURE	BCD Radio Parts	3
TERMINAL BLOCK: TELEPHONE TYPE	Fair Radio Sales	4
TERMINAL BOARD: KNURLED NUT POSTS	ETCO	2
TERMINAL BOARD: SCREW TYPE	ETCO	4
TERMINAL BOARD: SCREW TYPE	Mouser Electronics	14
TERMINAL BOARD: SPRING RETURN POSTS	ETCO	1
TERMINAL BOARD: TURRET POSTS	ETCO	7
TERMINAL STRIP SHORT: 2 LUG	Surplus Electronics	1
TERMINAL STRIP:	Mouser Electronics	7
TERMINAL STRIP:	Jameco Electronics	1
TERMINAL STRIP:	Surplus Electronics	2
TERMINAL STRIP:	Star-Tronics	3
TERMINAL STRIP:	Quest Electronics	4
TERMINAL STRIP:	ETCO	12
TERMINAL STRIP:	Fair Radio Sales	2
TERMINAL STRIP:	Daytapro Electronics	7
TERMINAL STRIP: BINDING POST, SCREW	Digi-Key	10
TERMINAL STRIP: MODULAR	Digi-Key	1
TERMINAL STRIP: PRINTED CKT BOARD	Digi-Key	3
TERMINAL STRIP: SCREW TYPE	Aldelco	1
TERMINAL: PERF-BOARD	Digi-Key	1
TERMINAL: PERF-BOARD	Circuit Specialists	3
TERMINAL: PERF-BOARD	Sintec	4
TERMINAL: PERF-BOARD	Mouser Electronics	25
TERMINAL: PERF-BOARD	ETCO	7
TERMINAL: PERF-BOARD, T42, T44	Circuit Specialists	2
TERMINAL: PERF-BOARD, VECTOR PIN	Active Electronics	4
TERMINAL: PERF-BOARD, VECTOR PIN	Jameco Electronics	7
TERMINAL: PERF-BOARD, VECTOR PIN	Quest Electronics	7
TERMINAL: PERF-BOARD, WIRE WRAP	Digi-Key	4
TERMINAL: PERF-BOARD, WIRE WRAP	Quest Electronics	4
TERMINAL: WIRE CONNECTOR	See Hardware: Terminal	
TERMINATION: RF	See Also Load:	
TERMINATION: RF, COAXIAL, OPEN CKT	Elcom Systems	1
TERMINATION: RF, COAXIAL, OPEN CKT	Lectronic	
TERMINATION: RF, COAXIAL, SHORT CKT	Elcom Systems	1
TERMINATION: RF, COAXIAL, SHORT CKT	Lectronic	
TERMINATION: RF, PILL, STRIPLINE	Lectronic	
TERMINATION: RF, WAVEGUIDE	Lectronic	
THERMAL COMPOUND:	See Chemical:	

COMPONENT	COMPANY NAME	VARIETY STOCKED
THERMISTER: 5K OHM @ 25 C, FENWAL	Digital Research	1
THERMISTER: 5K OHM @ 25 C, FENWAL	BCD Radio Parts	1
THERMISTER: 60.3 OHM @ 25 DEG C	ETCO	1
THERMISTER: TV	Fuji-Svea	6
TRANSFORMER:	Schnobel	
TRANSFORMER: AUDIO	Fair Radio Sales	8
TRANSFORMER: AUDIO	ETCO	17
TRANSFORMER: AUDIO	Semiconductors Surplus	2
TRANSFORMER: AUDIO, 10K TO 600 OHM	Star-Tronics	1
TRANSFORMER: AUDIO, INTER-STAGE	Surplus Electronics	3
TRANSFORMER: AUDIO, MINIATURE	Circuit Specialists	4
TRANSFORMER: AUDIO, MINIATURE	Mouser Electronics	94
TRANSFORMER: AUDIO, MINIATURE, PC	Fuji-Svea	2
TRANSFORMER: AUDIO, SUBMINIATURE	Fuji-Svea	10
TRANSFORMER: AUDIO, SUBMINIATURE	Circuit Specialists	2
TRANSFORMER: AUDIO, TRANSISTOR	Circuit Specialists	3
TRANSFORMER: AUDIO, ULTRA-MINIATURE	Fuji-Svea	5
TRANSFORMER: AUTO, STEP DOWN	Signal Transformer	22
TRANSFORMER: AUTO, STEP DOWN	ETCO	1
TRANSFORMER: BOBBIN, PLASTIC	Star-Tronics	1
TRANSFORMER: CLOCK TYPE, PC MOUNT	Jameco Electronics	4
TRANSFORMER: CONSTANT VOLTAGE	Fair Radio Sales	10
TRANSFORMER: CONSTANT VOLTAGE, 115V	ETCO	1
TRANSFORMER: CONSTANT VOLTAGE, 118V	Star-Tronics	1
TRANSFORMER: FLASH TUBE, HIGH VOLT	Mouser Electronics	4
TRANSFORMER: FLYBACK, DIGITAL EQUIP	Semiconductors Surplus	1
TRANSFORMER: FLYBACK, TV,THORDARSON	Fuji-Svea	54
TRANSFORMER: HIGH CURRENT, 3-PHASE	Signal Transformer	6
TRANSFORMER: HIGH VOLTAGE, FREQ	Star-Tronics	1
TRANSFORMER: HV, POWER SUPPLY	Fair Radio Sales	12
TRANSFORMER: HV, POWER SUPPLY	Amp Supply	11
TRANSFORMER: IF, RADIO	See Inductor: IF	
TRANSFORMER: INDUCTIVE, 2:1 RATIO	BCD Radio Parts	1
TRANSFORMER: INDUCTIVE, 2:1 RATIO	Digital Research	1
TRANSFORMER: ISOLATION, 120-120 VAC	Fair Radio Sales	1
TRANSFORMER: ISOLATION, 120-120 VAC	Fuji-Svea	1
TRANSFORMER: ISOLATION, POWER	Signal Transformer	16
TRANSFORMER: LOW VOLTAGE	BCD Radio Parts	9
TRANSFORMER: LOW VOLTAGE	Fair Radio Sales	78
TRANSFORMER: LOW VOLTAGE	Surplus Electronics	2
TRANSFORMER: LOW VOLTAGE	Quest Electronics	8
TRANSFORMER: LOW VOLTAGE	Semiconductors Surplus	22
TRANSFORMER: LOW VOLTAGE	Amp Supply	6
TRANSFORMER: LOW VOLTAGE	Digital Research	5
TRANSFORMER: LOW VOLTAGE	ETCO	23
TRANSFORMER: LOW VOLTAGE, CT	Digi-Key	24
TRANSFORMER: LOW VOLTAGE, FILAMENT	Jameco Electronics	11
TRANSFORMER: LOW VOLTAGE, FLATHEAD	Signal Transformer	23
TRANSFORMER: LOW VOLTAGE, HAMMOND	Daytapro Electronics	13
TRANSFORMER: LOW VOLTAGE, MINIATURE	Signal Transformer	50
TRANSFORMER: LOW VOLTAGE, MINIATURE	Star-Tronics	1
TRANSFORMER: LOW VOLTAGE, MOUSER	Mouser Electronics	29
TRANSFORMER: LOW VOLTAGE, MULTI-TAP	Mouser Electronics	1
TRANSFORMER: LOW VOLTAGE, P.C. MT	Daytapro Electronics	15
TRANSFORMER: LOW VOLTAGE, P.C. MT	Mouser Electronics	7
TRANSFORMER: LOW VOLTAGE, P.C. MT	Signal Transformer	60
TRANSFORMER: LOW VOLTAGE, WEIGHT	Signal Transformer	25
TRANSFORMER: LV & MV SECONDARY	Mouser Electronics	2

COMPONENT	COMPANY NAME	VARIETY STOCKED
TRANSFORMER: LV & MV SECONDARY	Star-Tronics	1
TRANSFORMER: LV, MV & HV SECONDARY	Star-Tronics	1
TRANSFORMER: LV, VERY HIGH CURRENT	Signal Transformer	11
TRANSFORMER: MICROPROCESSOR APPL.	Signal Transformer	8
TRANSFORMER: MODULATION	Fair Radio Sales	6
TRANSFORMER: MV, POWER SUPPLY	Fair Radio Sales	41
TRANSFORMER: RADIO IF	See Inductor: IF	
TRANSFORMER: RECTIFIER POWER	Signal Transformer	72
TRANSFORMER: RECTIFIER POWER	Circuit Specialists	8
TRANSFORMER: REGULATING, LOW VOLT.	Signal Transformer	28
TRANSFORMER: RF, COAX, 50 TO 75 OHM	Elcom Systems	1
TRANSFORMER: RF, COAX, 50 TO 93 OHM	Elcom Systems	1
TRANSFORMER: RF, WIDEBAND,TO .6 GHZ	Mini-Circuits Laboratory	40
TRANSFORMER: VARIABLE VOLTAGE	Fair Radio Sales	13
TRANSFORMER: VARIABLE, 0-120 VAC	Fuji-Svea	7
TRANSFORMER: VARIABLE, 0-120 VAC	Mouser Electronics	7
TRANSFORMER: VARIABLE, 0-130 VAC	ETCO	1
TRANSFORMER: WALL PLUG, LOW VOLTAGE	ETCO	4
TRANSFORMER: WALL PLUG, LOW VOLTAGE	Quest Electronics	9
TRANSFORMER: WALL PLUG, LOW VOLTAGE	BCD Radio Parts	1
TRANSFORMER: WALL PLUG, LOW VOLTAGE	Surplus Electronics	6
TRANSFORMER: WALL PLUG, LOW VOLTAGE	Jameco Electronics	6
TRANSFORMER: WALL PLUG, LOW VOLTAGE	Digital Research	3
TRANSFORMER: WALL PLUG, LOW VOLTAGE	Semiconductors Surplus	6
TRANSFORMER: WALL PLUG, TELEPHONE	ETCO	2
TRANSIENT SURGE ABSORBER: TRANSORB	Active Electronics	9
TRANSIENT SURGE ABSORBER: ZNR	Digi-Key	14
TRANSIENT SURGE ABSORBER: ZNR	Sintec	3
TRANSIENT VOLTAGE SUPRESSOR:	See Diode:	
TRANSISTOR:	Star-Tronics	10
TRANSISTOR: 2N6603, MICROWAVE	MHZ Electronics	1
TRANSISTOR: 2NXXXX SERIES	Digital Research	14
TRANSISTOR: 2NXXXX SERIES	Electronic Marketplace	7
TRANSISTOR: 2NXXXX SERIES	ETCO	23
TRANSISTOR: 2NXXXX SERIES	Key Electronics	5
TRANSISTOR: 2NXXXX SERIES	Circuit Specialists	225
TRANSISTOR: 2NXXXX SERIES	Fuji-Svea	169
TRANSISTOR: 2NXXXX SERIES	Semiconductors Surplus	55
TRANSISTOR: 2NXXXX SERIES	Digi-Key	13
TRANSISTOR: 2NXXXX SERIES	Babylon Electronics	52
TRANSISTOR: 2NXXXX SERIES	Aldelco	192
TRANSISTOR: 2NXXXX SERIES	Fair Radio Sales	3
TRANSISTOR: 2NXXXX SERIES	Jameco Electronics	44
TRANSISTOR: 2NXXXX SERIES	BCD Radio Parts	5
TRANSISTOR: 2NXXXX SERIES	Quest Electronics	68
TRANSISTOR: 2NXXXX SERIES	Radiokit	10
TRANSISTOR: 2NXXXX SERIES	Alpha Electronic Labs	16
TRANSISTOR: 2NXXXX SERIES	Active Electronics	249
TRANSISTOR: 2NXXXX SERIES, FET	Aldelco	8
TRANSISTOR: 2NXXXX SERIES, MOTOROLA	Mouser Electronics	188
TRANSISTOR: 2NXXXX SERIES, POWER	Sintec	30
TRANSISTOR: 2NXXXX SERIES, PWR TAB	Mouser Electronics	8
TRANSISTOR: 2NXXXX SERIES, SGS	Mouser Electronics	135
TRANSISTOR: 2NXXXX SERIES, THOMSON	Mouser Electronics	148
TRANSISTOR: 2SA- SERIES, JAPANESE	ETCO	7
TRANSISTOR: 2SA- SERIES, JAPANESE	Fuji-Svea	138

COMPONENT	COMPANY NAME	VARIETY STOCKED
TRANSISTOR: 2SB- SERIES, JAPANESE	Fuji-Svea	100
TRANSISTOR: 2SB- SERIES, JAPANESE	ETCO	5
TRANSISTOR: 2SC- SERIES, JAPANESE	Fuji-Svea	357
TRANSISTOR: 2SC- SERIES, JAPANESE	ETCO	3
TRANSISTOR: 2SD- SERIES, JAPANESE	Fuji-Svea	121
TRANSISTOR: 2SD- SERIES, JAPANESE	ETCO	2
TRANSISTOR: 2SK- SERIES, FET	Alpha Electronic Labs	1
TRANSISTOR: 2SK- SERIES, JAPANESE	Fuji-Svea	12
TRANSISTOR: 3NXXXX SERIES, FET	Alpha Electronic Labs	3
TRANSISTOR: 3NXXXX SERIES, FET	Mouser Electronics	1
TRANSISTOR: 3SK- SERIES, FET	Alpha Electronic Labs	1
TRANSISTOR: 3SK- SERIES, JAPANESE	Fuji-Svea	10
TRANSISTOR: 40327	Babylon Electronics	1
TRANSISTOR: 40347, RCA	Sintec	1
TRANSISTOR: 40374 WITH HEATSINK	Babylon Electronics	1
TRANSISTOR: 40409, 40410	Jameco Electronics	2
TRANSISTOR: 40673, DUAL GATE FET	Circuit Specialists	1
TRANSISTOR: 40673, DUAL GATE FET	Jameco Electronics	1
TRANSISTOR: 40673, DUAL GATE FET	Alaska Microwave Lab	1
TRANSISTOR: 40673, DUAL GATE FET	Aldelco	1
TRANSISTOR: 40673, DUAL GATE FET	Alpha Electronic Labs	1
TRANSISTOR: 40841	Alpha Electronic Labs	1
TRANSISTOR: 92PU- SERIES, AUDIO PWR	Digi-Key	4
TRANSISTOR: AC- SERIES	ETCO	4
TRANSISTOR: AD- SERIES	ETCO	2
TRANSISTOR: ADZ- SERIES	ETCO	2
TRANSISTOR: BC- SERIES	ETCO	4
TRANSISTOR: BF- SERIES	ETCO	1
TRANSISTOR: BFQ85, MICROWAVE	MHZ Electronics	1
TRANSISTOR: BFR90, MICROWAVE	Fuji-Svea	1
TRANSISTOR: BFR90, MICROWAVE	Circuit Specialists	1
TRANSISTOR: BFW- SERIES, MICROWAVE	Semiconductors Surplus	2
TRANSISTOR: BFW92, MICROWAVE	MHZ Electronics	1
TRANSISTOR: BLY38	Semiconductors Surplus	1
TRANSISTOR: BRF- SERIES, MICROWAVE	Semiconductors Surplus	3
TRANSISTOR: BRF- SERIES, MICROWAVE	Alpha Electronic Labs	2
TRANSISTOR: BRF- SERIES, MICROWAVE	Alaska Microwave Lab	2
TRANSISTOR: BU208, HORIZONTAL DEFL	Fuji-Svea	1
TRANSISTOR: BU208, HORIZONTAL DEFL	Circuit Specialists	1
TRANSISTOR: BU208, HORIZONTAL DEFL	Semiconductors Surplus	1
TRANSISTOR: C106- SERIES, SCR	Fuji-Svea	6
TRANSISTOR: CB 5-WATT FINAL	Circuit Specialists	3
TRANSISTOR: D- SERIES	Babylon Electronics	4
TRANSISTOR: D40, D44	Digital Research	2
TRANSISTOR: D44	BCD Radio Parts	1
TRANSISTOR: DIAC	Star-Tronics	1
TRANSISTOR: DIAC, TECCOR	Mouser Electronics	5
TRANSISTOR: ECG- SERIES	ETCO	1
TRANSISTOR: FET, RF	Circuit Specialists	27
TRANSISTOR: GE- SERIES	ETCO	2
TRANSISTOR: HEP- SERIES, MOTOROLA	ETCO	21
TRANSISTOR: IRTR- SERIES	ETCO	3
TRANSISTOR: J- SERIES, FET	Active Electronics	4
TRANSISTOR: J- SERIES, FET	Fuji-Svea	2
TRANSISTOR: MAC- SERIES, TRIAC	Fuji-Svea	5
TRANSISTOR: MBS4991, BIDIRECTIONAL	Circuit Specialists	1
TRANSISTOR: MBS4992, BIDIRECTIONAL	Fuji-Svea	1
TRANSISTOR: MCR- SERIES, SCR	Fuji-Svea	8
TRANSISTOR: MD8001 DUAL NPN	Circuit Specialists	1

COMPONENT	COMPANY NAME	VARIETY STOCKED
TRANSISTOR: MD8003 DUAL NPN	Circuit Specialists	1
TRANSISTOR: MEM- SERIES, FET	Semiconductors Surplus	2
TRANSISTOR: MFE- SERIES, MOS FET	Fuji-Svea	2
TRANSISTOR: MGF- , FET, MICROWAVE	Alaska Microwave Lab	3
TRANSISTOR: MICROWAVE POWER, HEW PA	MHZ Electronics	1
TRANSISTOR: MJ- SERIES	Babylon Electronics	3
TRANSISTOR: MJ- SERIES, MOTOROLA	Mouser Electronics	5
TRANSISTOR: MJ- SERIES, POWER	Fuji-Svea	16
TRANSISTOR: MJ- SERIES, POWER	Circuit Specialists	43
TRANSISTOR: MJ- SERIES, POWER	Digi-Key	1
TRANSISTOR: MJ- SERIES, POWER	Alpha Electronic Labs	1
TRANSISTOR: MJ- SERIES, POWER, RCA	Sintec	1
TRANSISTOR: MJ- SERIES, SGS	Mouser Electronics	17
TRANSISTOR: MJE- SERIES	Quest Electronics	4
TRANSISTOR: MJE- SERIES	Jameco Electronics	2
TRANSISTOR: MJE- SERIES	Babylon Electronics	3
TRANSISTOR: MJE- SERIES, POWER	Fuji-Svea	29
TRANSISTOR: MJE- SERIES, POWER	Digi-Key	12
TRANSISTOR: MJF3439, POWER	Digital Research	1
TRANSISTOR: MK 10	Fuji-Svea	1
TRANSISTOR: MM- SERIES	Fuji-Svea	8
TRANSISTOR: MMCM SERIES, MICROWAVE	Semiconductors Surplus	5
TRANSISTOR: MMT 70	Fuji-Svea	1
TRANSISTOR: MMT 74	Semiconductors Surplus	1
TRANSISTOR: MOUNTING KITS	Mouser Electronics	5
TRANSISTOR: MPF 102, JFET	Fuji-Svea	1
TRANSISTOR: MPF 102, JFET	Aldelco	1
TRANSISTOR: MPF 102, JFET	Quest Electronics	1
TRANSISTOR: MPF 102, JFET	Active Electronics	1
TRANSISTOR: MPF 102, JFET	Digi-Key	1
TRANSISTOR: MPF 102, JFET	Radiokit	1
TRANSISTOR: MPF 102, JFET	Daytapro Electronics	1
TRANSISTOR: MPF 102, JFET	Alpha Electronic Labs	1
TRANSISTOR: MPF 102, JFET	Circuit Specialists	1
TRANSISTOR: MPF 103, JFET	Active Electronics	1
TRANSISTOR: MPF 131, MOSFET	Digital Research	1
TRANSISTOR: MPF 131, MOSFET	BCD Radio Parts	1
TRANSISTOR: MPF 161, JFET	Fuji-Svea	1
TRANSISTOR: MPF 161, JFET	Radiokit	1
TRANSISTOR: MPL 01	Active Electronics	1
TRANSISTOR: MPQ2221, QUAD, PNP	Fuji-Svea	1
TRANSISTOR: MPQ2222, QUAD, NPN	Semiconductors Surplus	1
TRANSISTOR: MPS- SERIES	Jameco Electronics	6
TRANSISTOR: MPS- SERIES	Quest Electronics	4
TRANSISTOR: MPS- SERIES	Key Electronics	2
TRANSISTOR: MPS- SERIES	Active Electronics	40
TRANSISTOR: MPS- SERIES	Alpha Electronic Labs	2
TRANSISTOR: MPS- SERIES	Digital Research	2
TRANSISTOR: MPS- SERIES	Digi-Key	11
TRANSISTOR: MPS- SERIES	Babylon Electronics	2
TRANSISTOR: MPS- SERIES, MOTOROLA	Fuji-Svea	92
TRANSISTOR: MPS- SERIES, MOTOROLA	Circuit Specialists	88
TRANSISTOR: MPS- SERIES, MOTOROLA	Mouser Electronics	83
TRANSISTOR: MPS- SERIES, MOTOROLA	Fair Radio Sales	2
TRANSISTOR: MPSA- SERIES	Digi-Key	1
TRANSISTOR: MPSA- SERIES	Quest Electronics	6
TRANSISTOR: MPSA- SERIES	Babylon Electronics	2
TRANSISTOR: MPSA- SERIES	Active Electronics	19
TRANSISTOR: MPSA- SERIES	Jameco Electronics	2

COMPONENT	COMPANY NAME	VARIETY STOCKED
TRANSISTOR: MPSA- SERIES	Alpha Electronic Labs	4
TRANSISTOR: MPSA- SERIES, MOTOROLA	Mouser Electronics	6
TRANSISTOR: MPSA- SERIES, MOTOROLA	Fuji-Svea	17
TRANSISTOR: MPSA-14	Key Electronics	1
TRANSISTOR: MPSA05, 06, AUDIO	Mouser Electronics	2
TRANSISTOR: MPSD- SERIES	Fuji-Svea	2
TRANSISTOR: MPSH- SERIES	Fuji-Svea	2
TRANSISTOR: MPSL- SERIES	Fuji-Svea	2
TRANSISTOR: MPSU- SERIES	Digital Research	1
TRANSISTOR: MPSU- SERIES	Quest Electronics	1
TRANSISTOR: MPSU- SERIES	Babylon Electronics	2
TRANSISTOR: MPSU- SERIES, UNIWATT	Fuji-Svea	19
TRANSISTOR: MPSU- SERIES, UNIWATT	Circuit Specialists	18
TRANSISTOR: MPU131, THYRISTOR	Fuji-Svea	1
TRANSISTOR: MRD- SERIES	Fuji-Svea	5
TRANSISTOR: MRF- SERIES, MICROWAVE	Semiconductors Surplus	46
TRANSISTOR: MRF- SERIES, MICROWAVE	Alaska Microwave Lab	2
TRANSISTOR: MRF- SERIES, MICROWAVE	Aldelco	26
TRANSISTOR: MRF- SERIES, MICROWAVE	Fuji-Svea	22
TRANSISTOR: MRF454, 3-30 MHz, 80 W	Semiconductors Surplus	1
TRANSISTOR: MRF472	Radiokit	1
TRANSISTOR: MRF901, MICROWAVE	MHZ Electronics	
TRANSISTOR: MRF901, MICROWAVE	Alpha Electronic Labs	1
TRANSISTOR: MRF901, MICROWAVE	Semiconductors Surplus	1
TRANSISTOR: MRF901-904, MICROWAVE	Circuit Specialists	3
TRANSISTOR: MRO- SERIES	Mouser Electronics	FULL
TRANSISTOR: MWA SERIES, MICROWAVE	Semiconductors Surplus	6
TRANSISTOR: NE- , FET, MOS	California Eastern Lab	4
TRANSISTOR: NE- , LOW COST BIPOLAR	California Eastern Lab	33
TRANSISTOR: NE- , MICROWAVE BIPOLAR	California Eastern Lab	20
TRANSISTOR: NE- , MICROWAVE SWITCH	California Eastern Lab	24
TRANSISTOR: NE- , MICROWAVE, CHIP	California Eastern Lab	31
TRANSISTOR: NE- , MICROWV GaAs FET	California Eastern Lab	20
TRANSISTOR: NE- , MICRWV OSCILLATOR	California Eastern Lab	24
TRANSISTOR: NE- , MICRWV PWR BIPOLR	California Eastern Lab	38
TRANSISTOR: NE- , PNP	California Eastern Lab	10
TRANSISTOR: NE- , POWER BIPOLAR	California Eastern Lab	30
TRANSISTOR: NE- , POWER GaAs FET	California Eastern Lab	30
TRANSISTOR: NE- , UHF POWER BIPOLAR	California Eastern Lab	37
TRANSISTOR: NE- , VHF POWER BIPOLAR	California Eastern Lab	24
TRANSISTOR: NEC SERIES, MICROWAVE	Alaska Microwave Lab	3
TRANSISTOR: NEC SERIES, MICROWAVE	Alpha Electronic Labs	2
TRANSISTOR: NEO 2137 MICROWAVE, NEC	BCD Radio Parts	1
TRANSISTOR: NEX-,MICROWV OSCILLATOR	California Eastern Lab	4
TRANSISTOR: PMD 11K-60, DARLINGTON	Digital Research	1
TRANSISTOR: PN- SERIES	Jameco Electronics	8
TRANSISTOR: PN- SERIES	Active Electronics	16
TRANSISTOR: PN- SERIES	Quest Electronics	6
TRANSISTOR: POWER NPN	Daytapro Electronics	5
TRANSISTOR: POWER PNP	Daytapro Electronics	2
TRANSISTOR: RCA 105	Babylon Electronics	1
TRANSISTOR: REN- SERIES, JAPANESE	Fuji-Svea	268
TRANSISTOR: RF POWER, HAM	Circuit Specialists	11
TRANSISTOR: RF POWER, HF,VHF,UHF	Semiconductors Surplus	5
TRANSISTOR: SCR	Electronic Marketplace	2
TRANSISTOR: SCR	Jameco Electronics	7
TRANSISTOR: SCR	Daytapro Electronics	5
TRANSISTOR: SCR	Circuit Specialists	18
TRANSISTOR: SCR	ETCO	4

COMPONENT	COMPANY NAME	VARIETY STOCKED
TRANSISTOR: SCR	Key Electronics	1
TRANSISTOR: SCR	Fuji-Svea	19
TRANSISTOR: SCR	Babylon Electronics	1
TRANSISTOR: SCR	Digital Research	2
TRANSISTOR: SCR	Proverty Electronics	1
TRANSISTOR: SCR	BCD Radio Parts	3
TRANSISTOR: SCR, 2N5060	Semiconductors Surplus	1
TRANSISTOR: SCR, GE EXACT REPLACEMT	Mouser Electronics	12
TRANSISTOR: SCR, RCA	Sintec	1
TRANSISTOR: SCR, TECCOR	Mouser Electronics	117
TRANSISTOR: SD- SERIES	Aldelco	23
TRANSISTOR: SE- SERIES	ETCO	1
TRANSISTOR: SIDAC, TECCOR	Mouser Electronics	7
TRANSISTOR: SK- SERIES	ETCO	2
TRANSISTOR: SWITCHING, NPN, SMALL	Daytapro Electronics	6
TRANSISTOR: SWITCHING, PNP, SMALL	Daytapro Electronics	5
TRANSISTOR: THYRISTOR, UNIJUNCTION	Fuji-Svea	3
TRANSISTOR: TIC- SERIES, SCR	Active Electronics	9
TRANSISTOR: TIC- SERIES, THYRISTOR	BCD Radio Parts	2
TRANSISTOR: TIC- SERIES, TRIAC	Active Electronics	8
TRANSISTOR: TIL 81 PHOTO TRANSISTOR	Active Electronics	1
TRANSISTOR: TIP- SERIES	Mouser Electronics	48
TRANSISTOR: TIP- SERIES	Jameco Electronics	11
TRANSISTOR: TIP- SERIES	BCD Radio Parts	6
TRANSISTOR: TIP- SERIES	Quest Electronics	12
TRANSISTOR: TIP- SERIES	Babylon Electronics	2
TRANSISTOR: TIP- SERIES	ETCO	1
TRANSISTOR: TIP- SERIES	Aldelco	11
TRANSISTOR: TIP- SERIES	Fair Radio Sales	5
TRANSISTOR: TIP- SERIES	Alpha Electronic Labs	28
TRANSISTOR: TIP- SERIES, POWER	Circuit Specialists	20
TRANSISTOR: TIP- SERIES, POWER	Sintec	5
TRANSISTOR: TIP- SERIES, POWER	Digital Research	2
TRANSISTOR: TIP- SERIES, POWER	Mouser Electronics	4
TRANSISTOR: TIP- SERIES, POWER	Active Electronics	69
TRANSISTOR: TIP- SERIES, POWER	Digi-Key	13
TRANSISTOR: TIP-120	Proverty Electronics	1
TRANSISTOR: TIP-125	Electronic Marketplace	1
TRANSISTOR: TIP-50, 400 VOLT, 40 W	Star-Tronics	1
TRANSISTOR: TIS- SERIES	Active Electronics	12
TRANSISTOR: TIS- SERIES	Quest Electronics	4
TRANSISTOR: TIS- SERIES	Jameco Electronics	2
TRANSISTOR: TRIAC	Circuit Specialists	15
TRANSISTOR: TRIAC	Jameco Electronics	3
TRANSISTOR: TRIAC	ETCO	3
TRANSISTOR: TRIAC	BCD Radio Parts	3
TRANSISTOR: TRIAC	Babylon Electronics	6
TRANSISTOR: TRIAC	Electronic Marketplace	1
TRANSISTOR: TRIAC	Fuji-Svea	9
TRANSISTOR: TRIAC	Semiconductors Surplus	1
TRANSISTOR: TRIAC	Digital Research	4
TRANSISTOR: TRIAC, FASTPAK	Mouser Electronics	4
TRANSISTOR: TRIAC, INTERNAL DIAC	Mouser Electronics	16
TRANSISTOR: TRIAC, LOGIC, TECCOR	Mouser Electronics	26
TRANSISTOR: TRIAC, RCA	Sintec	5
TRANSISTOR: TRIAC, TECCOR	Mouser Electronics	48
TRANSISTOR: UNIJUNCTION	Circuit Specialists	3
TRANSISTOR: UPT- SERIES, POWER	Active Electronics	2
TRANSISTOR: VN- SERIES, FET	Active Electronics	2

COMPONENT	COMPANY NAME	VARIETY STOCKED
TRANSISTOR: VN- SERIES, VMOS FET	Active Electronics	5
TRANSISTOR: VQ 1000 VMOS ARRAY	Active Electronics	1
TRIAC:	See Transistor: TRIAC	
TUBE BASE: OCTAL, BASE SHELL	ETCO	1
TUBE: 7651, RCA, WITH SOCKET	Semiconductors Surplus	1
TUBE: CATHODE RAY, B-12, 12"MONITOR	Surplus Electronics	1
TUBE: CATHODE RAY, SCOPE TYPE	Fair Radio Sales	34
TUBE: CATHODE RAY, SCOPE TYPE	ETCO	7
TUBE: CATHODE RAY, TELEVISION	ETCO	21
TUBE: CATHODE RAY, VIDEO MONITOR	Semiconductors Surplus	1
TUBE: CHIMNEY, EIMAC	MHZ Electronics	9
TUBE: CHIMNEY, FOR 4X150,4CX250	Fair Radio Sales	1
TUBE: CUP, HIGH VOLTAGE	Fuji-Svea	1
TUBE: ELECTRON, RECEIVEING,U/UU/NEW	Fair Radio Sales	358
TUBE: ELECTRON, RECEIVING	ETCO	1785
TUBE: ELECTRON, RECEIVING	Fuji-Svea	165
TUBE: ELECTRON, TRANSMITTING	MHZ Electronics	141
TUBE: ELECTRON, TRANSMITTING	Semiconductors Surplus	31
TUBE: ELECTRON, TRANSMITTING	Amp Supply	14
TUBE: ELECTRON, TRANSMITTING, EIMAC	Fuji-Svea	117
TUBE: ELECTRON, TRANSMITTING, EIMAC	Amateur Electronic Supply	13
TUBE: ELECTRON, TRANSMITTING, U/UU	Fair Radio Sales	16
TUBE: FLASH, SYLVANIA	Fair Radio Sales	1
TUBE: FLASH, XENON GAS	Mouser Electronics	7
TUBE: INFRARED IMAGING CONVERTER	ETCO	1
TUBE: MAGNETRON, LESS EXT MAGNETS	Fair Radio Sales	7
TUBE: PLATE CAP, ALUMINUM HEAT SINK	Amp Supply	2
TUBE: PLATE CAP, CERAMIC	Semiconductors Surplus	2
TUBE: PLATE CAP, CERAMIC	Radiokit	2
TUBE: SHEILD, CATHODE RAY TUBE	Fair Radio Sales	3
TUBE: SHEILD, METAL, MINIATURE TUBE	Fair Radio Sales	1
TUBE: SOCKET	See Socket: Tube	
TUBE: SOLID STATE REPL, COLLINS EQP	Sartori Associates	1
TUBE: SOLID STATE REPL, DRAKE EQPT	Sartori Associates	6
TUBE: TWT, 8-9.6 GHz, 5 WATT	Fair Radio Sales	1
TUBE: TWT, ITT D-2013, 2-4 GHz, 1KW	Fair Radio Sales	1
TUBE: TWT, RCA 4012, 2-4 GHz, 20 mW	Fair Radio Sales	1
TUBE: VIDICON	ATV Research	16
TUBE: VIDICON, 8541 #274J-, RCA	Semiconductors Surplus	3
TUBING CLAMP: PLASTIC, 1/4 TO 1" OD	Small Parts	7
TUBING CLAMP: SS, 1/4 TO 5-5/8"O.D.	Small Parts	11
TUBING CONNECTOR: STAINLESS STEEL	Small Parts	39
TUBING: ALUM, RND, TELESCOPING, 3'L	Small Parts	8
TUBING: ALUMINUM, .01" TO 4" OD	Precision Tube Company	
TUBING: BRASS, .01" TO 3" OD	Precision Tube Company	
TUBING: BRASS, HEX, TELESCOPING,1'L	Small Parts	4
TUBING: BRASS, RECTANGULAR, 3' L	Small Parts	4
TUBING: BRASS, RND, TELESCOPING,3'L	Small Parts	20
TUBING: BRASS, ROUND, 3/4-3" OD	Small Parts	6
TUBING: BRASS, SQ, HEAVY WALL, 3' L	Small Parts	9
TUBING: BRASS, SQ, TELESCOPING, 3'L	Small Parts	7
TUBING: BRONZE, .01" TO 3" OD	Precision Tube Company	
TUBING: COPPER, .01" TO 3" OD	Precision Tube Company	

COMPONENT	COMPANY NAME	VARIETY STOCKED
TUBING: COPPER, RND, 1/8 TO 1.5" OD	Small Parts	9
TUBING: COPPER, RND,TELESCOPING,3'L	Small Parts	4
TUBING: HEAT SHRINK	Star-Tronics	4
TUBING: HEAT SHRINK	Mouser Electronics	17
TUBING: HEAT SHRINK	Daytapro Electronics	1
TUBING: HEAT SHRINK	Quest Electronics	1
TUBING: HEAT SHRINK	ETCO	2
TUBING: HEAT SHRINK	Active Electronics	
TUBING: HEAT SHRINK, ADHESIVE WALL	Digi-Key	6
TUBING: HEAT SHRINK, ASSORTMENT	Fuji-Svea	1
TUBING: HEAT SHRINK, POLYOLEFIN	Digi-Key	14
TUBING: HEAT SHRINK, TEFLON	Small Parts	18
TUBING: PLASTIC, CLEAR, 1/8-1/2" ID	Small Parts	4
TUBING: SPAGHETTI, ASSORTMENT	Fuji-Svea	1
TUBING: SPAGHETTI, SILICONE RUBBER	Star-Tronics	2
TUBING: SPAGHETTI, TEFLON	Frank Wirt Electronics & TV	1
TUBING: STAINLESS STEEL, 1/8-1" OD	Small Parts	6
TUBING: STAINLESS STEEL, CAPILLARY	Small Parts	44
TUBING: TEFLON	Small Parts	20
TUBING: VINYL, 7/8" DIA	ETCO	1
TUNING FORK: 87.6, 96.19, 180 VPS	Typetronics	3
TUNING FORK: 96.19, 120 VPS	Atlantic Surplus Sales	2
VALVE: AIR, LOW PRESSURE, 12 VDC	ETCO	1
VALVE: LIQUID, AIR, 12 VDC SOLENOID	ETCO	1
VARISTOR:	See Diode: Transient Voltage	
VELCRO: STRIP	BCD Radio Parts	1
VELCRO: STRIP	Digital Research	1
VIDICON:	See Tube: Vidicon & Part #	
WARMING PLATE: 117 V, TEMP CONTROL	Semiconductors Surplus	1
WAVEGUIDE: BULKHEAD FEEDTHRU	Lectronic	
WAVEGUIDE: COAXIAL ADAPTER	Lectronic	
WAVEGUIDE: FLANGE ADAPTER	Lectronic	
WAVEGUIDE: JOINT, ROTATING	Lectronic	
WAVEGUIDE: MODE ABSORBER	Lectronic	
WAVEGUIDE: MOUNT, BOLOMETER	Lectronic	
WAVEGUIDE: MOUNT, DETECTOR	Lectronic	
WAVEGUIDE: MOUNT, THERMISTER	Lectronic	
WAVEGUIDE: SECTION, PRESSURIZING	Lectronic	
WAVEGUIDE: SECTION, STRAIGHT	Lectronic	
WAVEGUIDE: SECTION, TAPERED	Lectronic	
WAVEGUIDE: SECTION, TWIST, BEND	Lectronic	
WAVEGUIDE: SERIES TEE	Lectronic	
WAVEGUIDE: SHUNT TEE	Lectronic	
WAVEGUIDE: SHUTTER	Lectronic	
WAVEGUIDE: TRANSITION, KLYSTRON	Lectronic	
WAVEGUIDE: TRANSITION, MAGNETRON	Lectronic	
WAVEGUIDE: TRANSITION, TAPER	Lectronic	
WAVEGUIDE: TUNER, SLIDE SCREW	Lectronic	
WAVEGUIDE: WINDOW, PRESSURE	Lectronic	
WAVEGUIDE: WR-90 FLANGE	Alaska Microwave Lab	1
WAVEMETER:	See Also Frequency Meter:	

COMPONENT	COMPANY NAME	VARIETY STOCKED
WAVEMETER: CAVITY, WAVEGUIDE	Lectronic	
WIRE:	Schnobel	2
WIRE: ANNEALED, DARK	Small Parts	1
WIRE: ANTENNA	Radiokit	1
WIRE: ANTENNA RADIAL, ALUMINUM,	Van Gorden Engineering	1
WIRE: ANTENNA, #14x7 STRAND	Viking Instruments	1
WIRE: ANTENNA, COPPERWELD	Lacue Communications	3
WIRE: ANTENNA, COPPERWELD	Certified International	1
WIRE: ANTENNA, COPPERWELD	Barker & Williamson	1
WIRE: ANTENNA, COPPERWELD, STRANDED	G & C Communications	2
WIRE: ANTENNA, COPPERWELD, STRANDED	Lacue Communications	1
WIRE: ANTENNA, FORMVAR	Skylane Products	1
WIRE: ANTENNA, INDOOR	Amateur Electronic Supply	1
WIRE: ANTENNA, SOLID	Amateur Electronic Supply	2
WIRE: ANTENNA, STRANDED	Amateur Electronic Supply	1
WIRE: ANTENNA, STRANDED COPPER,14Ga	Lacue Communications	1
WIRE: BRAID	See Cable:	
WIRE: BRASS, SOFT	Small Parts	4
WIRE: BUS, SOLID, TINNED COPPER	Mouser Electronics	1
WIRE: COPPER, BARE	Small Parts	4
WIRE: GROUNDING, ALUMINUM, 8 Ga	Van Gorden Engineering	1
WIRE: GROUNDING, ALUMINUM, 8 Ga	Lacue Communications	1
WIRE: GROUNDING, ALUMINUM, 8 Ga	Amateur Electronic Supply	1
WIRE: GUY, 6-18 STEEL	Certified International	
WIRE: HOOK UP	Certified International	
WIRE: HOOK UP	Mouser Electronics	12
WIRE: HOOK UP	Semiconductors Surplus	1
WIRE: HOOK UP	Quest Electronics	3
WIRE: HOOK UP, PVC INSULATION	Star-Tronics	1
WIRE: HOOK UP, PVC INSULATION	Star-Tronics	1
WIRE: HOOK UP, SHEILDED	Star-Tronics	1
WIRE: HOOK UP, SHEILDED	Nemal Electronics	1
WIRE: HOOK UP, SHEILDED	Fair Radio Sales	1
WIRE: HOOK UP, SOLID	Mouser Electronics	6
WIRE: HOOK UP, SOLID	Sintec	5
WIRE: HOOK UP, SOLID	Fair Radio Sales	2
WIRE: HOOK UP, SOLID	Fuji-Svea	5
WIRE: HOOK UP, SOLID, PVC INSUL	Digi-Key	16
WIRE: HOOK UP, STRANDED	Fuji-Svea	5
WIRE: HOOK UP, STRANDED	Fair Radio Sales	1
WIRE: HOOK UP, STRANDED	Jameco Electronics	8
WIRE: HOOK UP, STRANDED	Sintec	5
WIRE: HOOK UP, STRANDED, PVC INSUL	Digi-Key	40
WIRE: HOOK UP, STRANDED, PVC INSUL	Digi-Key	2
WIRE: MAGNET	Mouser Electronics	10
WIRE: MAGNET	Amidon Associates	11
WIRE: MAGNET	Radiokit	
WIRE: MAGNET	Semiconductors Surplus	5
WIRE: MAGNET	Star-Tronics	2
WIRE: MUSIC, STEEL	Small Parts	10
WIRE: NUTS	Mouser Electronics	3
WIRE: NYLON INSULATION, SHEILDED	Fair Radio Sales	1
WIRE: PRE-CUT JUMPER, .5-4"L, 20 Ga	Sintec	1
WIRE: RIBBON CABLE	See Cable: Ribbon	
WIRE: SILVERED, 14 Ga, TEFLON INSUL	Radiokit	
WIRE: SOLDER THRU'	Active Electronics	2
WIRE: SPEAKER	Fuji-Svea	2
WIRE: SPEAKER	Certified International	

COMPONENT	COMPANY NAME	VARIETY STOCKED
WIRE: SPEAKER, #18, 21, 24 Ga	ETCO	3
WIRE: SPEAKER, STRANDED	Mouser Electronics	3
WIRE: STAINLESS STEEL	Small Parts	17
WIRE: STEEL, GALVANIZED	Small Parts	5
WIRE: STEEL, PLASTIC COATED	Small Parts	1
WIRE: TEST LEAD	Mouser Electronics	8
WIRE: TEST LEAD, SILICONE RUBBER	Star-Tronics	1
WIRE: TWIST TIE, PLASTIC COATED	Small Parts	1
WIRE: WIRE WRAP	Quest Electronics	2
WIRE: WIRE WRAP	Semiconductors Surplus	3
WIRE: WIRE WRAP	Fuji-Svea	12
WIRE: WIRE WRAP, KYNAR	Jameco Electronics	6
WIRE: WIRE WRAP, KYNAR	Jameco Electronics	5
WIRE: WIRE WRAP, KYNAR	Sintec	5
WIRE: WIRE WRAP, KYNAR	Digi-Key	27
WIRE: WIRE WRAP, POLY SLIT	Active Electronics	1
WIRE: WIRE WRAP, PRE-CUT LENGTH	Sintec	24
WIRE: WIRE WRAP, PRE-CUT LENGTH	Digi-Key	24
WIRE: WIRE WRAP, PRE-CUT LENGTH	Fuji-Svea	24
WIRE: WIRE WRAP, PRE-CUT LENGTH	Jameco Electronics	ASST
WIRE: WIRE WRAP, PRE-CUT LENGTH	Quest Electronics	FULL
WIRE: WIRE WRAP, TEFZEL	Mouser Electronics	5
WIRE: WIRE WRAP, TEFZEL	Sintec	4
WIRE: WIRE WRAP, TEFZEL	Star-Tronics	1
WIRE: WIRE WRAP, TEFZEL SLIT	Active Electronics	1
WIRE: ZIP CORD	Mouser Electronics	6
YOKE: CRT, TV, COLOR	ETCO	11
YOKE: CRT, TV, THORDARSON	Fuji-Svea	86
ZENER DIODE:	See Diode: Zener	

SUPPLIER INFORMATION

The businesses listed in this section have indicated to us that they are willing to sell components to individuals in small quantities by mail. This list does not mean that these firms are in any way approved by us.

Included is the mailing address for each firm in the directory. Their phone number is included when available. Many of the firms listed will send their catalog free of charge, but some charge a small amount. If there is a fee, the amount is indicated under their business name. Please be considerate and include this amount when requesting a catalog. When a SASE is shown, they require that you send a self-addressed and stamped envelope for the return catalog.

Delays can be lessened by phoning the firm and asking about a specific part. When ordering, delays can be lessened by avoiding the use of personal checks, which can take as long as a week to clear.

Abbreviations are used under the company names with the following meanings:

(CC)	Catalog Cost. The amount to be paid when requesting a catalog.
(MO)	Minimum Order. The minimum order amount accepted.
(Sur)	Surplus. The parts sold by this company are industrial and/or military surplus.

ACTIVE ELECTRONIC SALES CORP.
(Free CC, $10.00 MO)
P.O. Box 8000
Westborough,
MA 01581 (617) 366-0500

ALASKA MICROWAVE LAB.
(Free CC, $0.00 MO)
4335 E. 5th Ave.
Anchorage,
AK 99504 (907) 338-0340

ALDELCO ELECTRONICS COMPANY
(Stamp CC, $0.00 MO)
2789 Milburn Ave.
Baldwin,
NY 11510 (516) 378-4555

ALPHA ELECTRONIC LABORATORIES
(Free CC, $0.00 MO)
2302 Oakland Gravel Road
Columbia,
MO 65202 (314) 874-1514

AMATEUR ELECTRONIC SUPPLY
(Free CC, $0.00 MO)
4828 W. Fond du Lac Ave.
Milwaukee,
WI 53216 (414) 442-4200

AMATEUR RADIO COMPONENTS SER.
(SASE CC, $0.00 MO)
P.O. Box 546 35 Highlands Dr.
East Greenbush,
NY 12061 (518) 477-4990

AMIDON ASSOCIATES, INC.
(Free CC, $0.00 MO)
12033 Otsego St.
North Hollywood,
CA 91607 (213) 760-4429

AMP SUPPLY CO.
(Free CC, $0.00 MO)
73 Maple Drive
Hudson,
OH 44236 (216) 656-4364

ATLANTIC SURPLUS SALES
(Free CC, $0.00 MO)
3730 Nautilus Avenue
Brooklyn,
NY 11224 (212) 372-0349

ATV RESEARCH
(Free CC, $0.00 MO)
13th & Broadway
Dakota City,
NE 68731 (402) 987-3771

B.G. MICRO
(Free CC, $0.00 MO)
P.O. Box 280298
Dallas,
TX 75228 (214) 271-5546

BABYLON ELECTRONICS
(Free CC, $0.00 MO)
P.O. Box 41778
Sacramento,
CA 95841 (916) 334-2161

BARKER & WILLIAMSON, INC.
(Free CC, $15.00 MO)
10 Canal St.
Bristol,
PA 19007 (215) 788-5581

BCD RADIO PARTS CO.
(Free CC, $0.00 MO, Sur)
P.O. Box 119
Richardson,
TX 75080-0020 (214) 238-0040

C-W CRYSTALS
(Free CC, $0.00 MO)
570 N. Buffalo St.
Marshfield,
MO 65706

CALIFORNIA EASTERN LABS, INC
(Free CC, $0.00 MO)
3005 Democracy Way
Santa Clara,
CA 95050 (408) 988-3500

CAYWOOD ELECTRONICS, INC.
(Free CC, $0.00 MO)
P.O. Drawer U
Malden,
MA 02148-0921

CERTIFIED INTERNATIONAL
(Free CC, $0.00 MO)
4138 So. Ferris
Fremont,
MI 49412 (616) 924-4561

CIRCUIT SPECIALISTS CO.
($1.00 CC, $0.00 MO)
P.O. Box 3047
Scottsdale,
AZ 85257 (602) 966-0764

CURTIS ELECTRO DEVICES, INC.
(Free CC, $0.00 MO)
Box 4090
Mountain View,
CA 94040 (415) 964-3846

D&V RADIO PARTS
(Stamp CC, $0.00 MO)
12805 W. Sarle, R#2
Freeland,
MI 48623 (517) 695-2210

DAYTAPRO ELECTRONICS, INC.
(Free CC, $0.00 MO)
3029 Wilshire Ln.
Arlington Hts.,
IL 60004 (312) 870-0555

DIGI-KEY CORPERATION
(Free CC, $0.00 MO)
Highway 32 South, PO Box 677
Thief River Falls,
MN 56701 (218) 681-6674

DIGITAL RESEARCH: PARTS
(Free CC, $0.00 MO)
P.O. Box 401247
Garland,
TX 75040 (214) 271-2461

DYNACLAD INDUSTRIES
($1.50 CC, $0.00 MO)
P.O. Box 296
Meadow Lands,
PA 15347 (412) 225-7082

EDMUND SCIENTIFIC (Free CC, $0.00 MO)	101 East Gloucester Pike Barrington, NJ	08007	(609) 547-3488
ELCOM SYSTEMS INC. (Free CC, $0.00 MO)	4032 Clint Moore Rd. Boca Raton, FL	33431	(305) 994-1774
ELECTRIC MOTION COMPANY (Free CC, $0.00 MO)	100 Whiting Street Winsted, CT	06098	(203) 379-8503
ELECTRONIC MARKETPLACE (Free CC, $0.00 MO)	5344 Jackson Drive La Mesa CA	92041	(714) 698-1777
ETCO ELECTRONICS (Free CC, $10.00 MO, Sur)	North County Shopping Center, Rt 9 N Plattsburgh, NY	12901	(518) 561-8700
FAIR RADIO SALES CO. (Free CC, $5.00 MO, Sur)	P.O. Box 1105, 1016 E. Eureka St. Lima, OH	45802	(419) 233-2196
FOX TANGO CORPORATION (Free CC, $0.00 MO)	P.O. Box 15944 West Plam Beach, FL	33406	(305) 683-9587
FRANK WIRT ELECTRONICS & TV (SASE CC, $0.00 MO)	Route 150 Alpha, IL	61413	(309) 529-5901
FUJI-SVEA (Free CC, $0.00 MO)	P.O. Box 3375 Torrance, CA	90510	(213) 533-1221
G & C COMMUNICATIONS (Free CC, $0.00 MO)	P.O. Box 5632, 1529 N. Cotner Blvd. Lincoln, NE	68505	(402) 464-1487
HAM HARDWARE HEADQUARTERS (Free CC, $3.00 MO)	29716 Briarbank Ct. Southfield, MI	48034	(313) 353-7582
INTERNATIONAL CRYSTAL MFG. CO. (Free CC, $0.00 MO)	10 N. Lee Oklahoma City OK	73102	(405) 236-3741
ITT MICROSYSTEMS DIVISION (Free CC, $0.00 MO)	Hillsboro Plaza, 700 N.W. 12th Avenue Deerfield Beach, FL	33441	(305) 421-8450
JAMECO ELECTRONICS (Free CC, $10.00 MO)	1355 Shoreway Road Belmont, CA	94002	(415) 592-8097

KEPRO CIRCUIT SYSTEMS, INC. (Free CC, $15.00 MO)	630 Axminister Drive Fenton, MO 63026-2992	(314) 343-1630
KEY ELECTRONICS (Free CC, $0.00 MO)	P.O. Box 3506 Schnectady, NY 12303	
LACUE COMMUNICATIONS, ELECT. (Free CC, $0.00 MO)	102 Village St. Johnstown, PA 15902	(814) 536-5500
LECTRONIC RESEARCH LABS, INC. (Free CC, $25.00 MO)	Atlantic & Ferry Ave. Camden, NJ 08104	(609) 541-4200
MHz ELECTRONICS (Free CC, $0.00 MO)	2111 W. Camelback Rd. Phoenix, AZ 85015	(602) 242-3037
MINI-CIRCUITS LABORATORY (Free CC, $0.00 MO)	2625 East 14th Street Brooklyn, NY 11235	(212) 769-0200
MJF ENTERPRISES, INC (Free CC, $0.00 MO)	P.O. Box 494 Mississippi State, MS 39762	(601) 323-5869
MOUSER ELECTRONICS (Free CC, $0.00 MO)	11433 Woodside Ave. Santee, CA 92071	(619) 449-2222
NEMAL ELECTRONICS CO. (Free CC, $0.00 MO)	5685 S.W. 80 St. Miami, FL 33143	(305) 661-5534
OPTOELECTRONICS (Free CC, $0.00 MO)	5821 N.E. 14th Avenue Ft. Lauderdale, FL 33334	(305) 771-2050
P. C. ELECTRONICS (Free CC, $0.00 MO)	2522 Paxson Lane Arcadia, CA 91006	(213) 447-4565
PALOMAR ENGINEERS (Free CC, $0.00 MO)	1924-F W. Mission Rd. Escondido, CA 92025	(619) 747-3343
PIEZO TECHNOLOGY INC. (Free CC, $0.00 MO)	2525 Shader Road, P.O. Box 7859 Orlando, FL 32854-7859	(305) 298-2000

PROVERTY ELECTRONICS
(Free CC, $0.00 MO)
R.R. Box 82
Edwards
IL 61528 (309) 692-6767

PRECISION TUBE COMPANY, INC.
(Free CC, $0.00 MO)
Church Road & Wissahickon Ave.
North Wales,
PA 19454 (215) 699-5801

QUEST ELECTRONICS
(Free CC, $5.00 MO)
P.O. Box 4430
Santa Clara,
CA 95054 (408) 988-1640

RADIOKIT
($0.50 CC, $0.00 MO)
Box 411
Greenville,
NH 03048 (603) 878-1033

RIW PRODUCTS
(Free CC, $0.00 MO)
Box 191
Babylon,
NY 11702

ROLIN DISTRIBUTORS
(Free CC, $0.00 MO)
P.O. Box 436
Dunellen,
NJ 08812 (201) 469-1219

SARTORI ASSOCIATES
(Free CC, $0.00 MO)
P.O. BOX 2085
Richardson,
TX 75080 (214) 494-3093

SAVOY ELECTRONICS INC.
(Free CC, $0.00 MO)
1175 N.E. 24 Street, P.O. Box 5727
Fort Lauderdale,
FL 33310 (305) 563-1333

SCHNOBEL
(Free CC, $0.00 MO)
General Delivery
Winnett,
MT 59087

SEMICONDUCTORS SURPLUS
(Free CC, $10.00 MO)
2822 No. 32nd St. Unit 1
Phoenix,
AZ 85008 (602) 956-9423

SENTRY MANUFACTURING COMPANY
(Free CC, $0.00 MO)
Crystal Park
Chickasha,
OK 73018 (405) 224-6780

SHERWOOD ENGINEERING INC.
(Free CC, $0.00 MO)
1268 S. Ogden St.
Denver,
CO 80210 (303) 722-2257

SIGNAL TRANSFORMER
(Free CC, $0.00 MO)
500 Bayview Ave.
Inwood,
NY 11696 (516) 239-4510

SINTEC ELECTRONICS (Free CC, $0.00 MO)	Drawer Q Milford, NJ 08848-9990	(201) 996-4093
SKYLANE PRODUCTS ($0.50 CC, $0.00 MO)	406 Bon-Aire Drive Temple Terrace, FL 33617	(813) 988-4213
SMALL PARTS INC. (Free CC, $0.00 MO)	6901 N.E. Third Ave., P.O. Box 381736 Miami, FL 33138	(305) 751-0856
SPECTRUM INTERNATIONAL, INC. (Free CC, $0.00 MO)	P.O. Box 1084 Concord, MA 01742	(617) 263-2145
STAR-TRONICS (Free CC, $4.00 MO, Sur)	P.O. BOX 683 McMinnville, OR 97128	(503) 472-9716
SURPLUS ELECTRONICS CORP. (Free CC, $15.00 MO, Sur)	7294 N.W. 54th Street Miami, FL 33166	(305) 887-8228
TEN-TEC INC (Free CC, $0.00 MO)	Highway 411 East Sevierville, TN 37862	(615) 453-7172
TYPETRONICS (SASE CC, $0.00 MO)	Box 8873 Ft. Lauderdale, FL 33310	(305) 583-1340
VAN GORDEN ENGINEERING (Free CC, $0.00 MO)	P.O. Box 21305 South Euclid, OH 44121	(216) 382-8410
VIKING INSTRUMENTS, INC. (Free CC, $0.00 MO)	73 Ferry Road Chester, CN 06412	(203) 526-5324